Animal Cell Culture and Virology

Animal Cell Culture and Virology

By :
S. NANDI
Senior Scientist
Centre for Animal Diseases Research and Diagnosis
Indian Veterinary Research Institute
Izatnagar - 243 122 (Uttar Pradesh) India

2009
New India Publishing Agency
Pitam Pura, New Delhi- 110 088

Published by
Sumit Pal Jain *for*

New India Publishing Agency
101, Vikas Surya Plaza, CU Block, L.S.C. Mkt.,
Pitam Pura, New Delhi- 110 088, (India)
Phone: 011-27341717, Fax: 011-27341616
E-mail: newindiapublishingagency@gmail.com
Web: www.bookfactoryindia.com

© Author: **2009**

All rights reserved, no part of this publication may be reproduced, stored in a retrieval system or transmitted in any form or by any means, electronic, mechanical, photocopying, recording or otherwise without the prior written permission of the publisher / author.

ISBN : 978-93-80235-05-9

Typeset at: Typographiya # 98 11 48 23 28
Printed at: Jai Bharat Printing Press, Delhi

Dedicated to the

Almighty

for everything

Preface

Cell culture is an essential and indispensable tool in many branches of the life sciences and the applications of cell culture is getting increased exponentially in various fields of biological and medical research arena. It provides the basics for studying the regulation of cell proliferation, differentiation and product formation. Viruses replicate only within the living cells and with the advent of cell culture technique, it has become easier to grow viruses under *in vitro* condition. Besides due to agitation shown by the animal activist groups, religious people and animal ethics committee members, testing of different metals (arsenic, cadmium, mercury etc), toxic substances, toxins, drugs, medicines, viruses etc in animals have become a nightmare. The use of cell culture made it easier for testing of above mentioned substances. This book entitled "Animal Cell Culture and Virology" is intended to explain why and how the basic techniques are used and how to set up a cell culture laboratory and detailed procedures of different techniques of cell culture starting from media filtration, sterilization of glasswares, plasticwares etc, counting of cells, cryopreservation of cells and revival, primary cell culture, secondary cell culture, contamination and curing, applications, advantages and disadvantages of cell culture, isolation and identification of viruses in cell culture, SNT, titration of viruses, scaling up of cell culture, collection, preservation and dispatch of samples for diagnosis of viral diseases, cell line authentication and characterization etc have been described in a simple and easily understandable manner. Lastly it will be useful not only for undergraduates, graduates, technicians, researchers but also teachers and scientists directly and indirectly associated with cell culture in day to day experiments. In spite of my best efforts at perfection, element of human error is still likely to creep, the author will appreciate receiving any comments on the improvement of the book.

Author

Content

Preface vii

CHAPTER
1. Introduction 1
2. Principles of Cell Culture 5
3. The Cell and its Environment 11
4. Types of Cells and Cell Culture 15
5. Cell Culture : Applications, Advantages and Disadvantages 21
6. Ideal Cell Culture Laboratory with Equipments ... 27
7. Growth and Maintenance of Cells in Culture .. 41
8. Components of Medium and its Preparation ... 47
9. Primary Cell Culture 57
10. Maintenance of Cell Culture 65
11. Packaging and Transportation of Cells 71
12. Description of Commonly Used Cell Lines 73
13. Subculturing of Cells in Different Vessels 79

14. Scaling Up of Animal Cell Culture 83
15. Cell Line Authentication and Characterization 89
16. Cryopreservation and Thawing/Revival of Cells 93
17. Contamination of Cell Culture and Curing 99
18. Counting of Cells 103
19. Isolation and Identification of Virus in Cell Culture 107
20. Virus Assays 111
21. Collection, Preservation and Transport of Various Specimens for Laboratory Diagnosis 115
22. Collection of Materials for Diagnosis of Viral Diseases 121
23. Processing of Laboratory Specimens, Inoculation into Laboratory Animals and Embryonated Eggs for Pathogen Isolation 127
24. Serum Neutralization Test 133
25. Enzyme Linked Immunosorbent Assay (ELISA) 139
26. Biosafety Measures and Risk Management 149

Glossary 153
Appendix 157
References 163

1
INTRODUCTION

Cell culture has become one of the major tools used in the life sciences today. Viruses replicate only within the living cells. With the advent of cell culture technique, it has become easier to grow the viruses under *in vitro* condition (cell culture). Viruses are usually propagated in cell culture, embryonated hen's egg, laboratory animals or natural hosts. In 1949, Enders, Weller and Robbins first reported that polio virus could be grown in cultured non-neural cells with the production of recognizable cytopathic changes. Although the cell culture technique has been evolved long back, it is only since the advent of antibiotics that cell culture became a matter of simple routine technique. Aseptic precautions are still essential, but the problems of contamination with bacteria, mycoplasma, fungi and yeasts can be checked by antibiotics and anti-fungal agents. The term *in vitro* literally means 'in glass' although today most cell culture is performed in or on plastic.

2 | Animal Cell Culture and Virology

Cell culture has been used in biology since the beginning of this century. It has been mostly used in the virology laboratory to isolate the virus, study the replication kinetics of virus, prepare large quantity of virus and prepare a number of vaccines of virus origin. The large scale production of virus has made it easier to characterize the protein and genomic nucleic acid of the virus. However, cell cultures of various sorts are considerably used in toxicology to replace the use of animals, supplement the use of animals and perhaps provide information that can be obtained by no other convenient means. The application of cell culture to toxicity testing of a number of chemicals, drugs, compounds etc has been of increasing interest over recent years. A variety of *in vitro* systems based on the primary, secondary or established cell lines are available now-a-days for toxicological research. The *in vitro* assays are generally easier to manipulate, less expensive to maintain, less complex to interpret, engender fewer ethical constraints, more reproducible and can give answers far quicker than that of laboratory animals.

Cell culture requires less space, is less expensive and is more convenient than the use of animals or eggs, although eggs are still the method of choice in the large scale preparation of certain vaccines. It is indispensable for the primary isolation of virus, infectivity assays, biochemical studies and production of vaccine. An additional advantage of tissue culture systems is that they are completely free from certain factors which restricts virus multiplication in the intact animal i.e virus specific antibodies and non-specific inhibitors of several kinds. Reproducible infection of a particular kind of cell by a given virus is thereby facilitated whereas animal may show various responses to challenge with a standard virus dose. Finally tissue culture is much more amenable to aseptic technique than those involving animals and eggs.

Although animal cell culture was first successfully undertaken by Ross Harrison in 1907, it was not until the late

1940's to early 1950's that several developments occurred that made cell culture widely available as a tool for scientist. First, there was the development of antibiotics that made it easier to avoid many of the contamination problems that earlier cell culture attempts faced. Second was the development of the techniques, such as the use of trypsin to remove cells from culture vessels, necessary to obtain continuously growing cell lines. Third, using these cell lines, scientist were able to develop, standardized, chemically defined culture media that made it far easier to grow cells. These three areas combined have allowed many more scientist to use cell, tissue and organ culture in their research. During the 1960's and 1970's, commercialization of this technology had further impact on cell culture that continues to this day. Companies such as Corning, Sigma, Invitrogen, Life Technologies etc began to develop and sell disposable plastic and glass cell culture products, improved filtration products and materials, liquid and powdered tissue culture media and laminar flow hoods. The overall result of these and other continuing technological developments has been a widespread increase in the number of laboratories and industries using cell culture today.

2
PRINCIPLES OF CELL CULTURE

The cell is the unit of structure of all plants and animals. With increasing complexity of organization, masses of different cell types tend to become localized and to form recognizable patterns. An aggregation of cells forming a definite pattern is called tissue. A still higher level of organization, tissues aggregate to form organs. The animal cell is a highly organized structure. Plant cell is different from animal cell by having a cellulose cell wall. There are a number of well defined structures viz., nucleus, mitochondria, lysosomes, Golgi apparatus etc in the cell. Animal cells also have centrioles whereas plant cells contain plastids such as chloroplasts. The remainder of the cell is the cytoplasm.

Cell types : On the basis of morphological and metabolic characteristics cell can be grouped into differences types. In the tissue culture, these differences tend to disappear. All the cells in an organism are derived from a single cell, fertilized egg. In the early stage, the embryo consists of ectoderm, mesoderm and endoderm.

Ectoderm gives rise to epithelia covering outer surface of the body including the mouth and nasal passages lining, nervous system, smooth muscle of the iris, anterior lobe of pituitary etc. The mesoderm gives rise to most of the body muscles, connective tissues, blood, endothelial lining of body cavity and urogenital system. The endoderm gives rise to all the epithelial lining of G.I. tract, respiratory tract, lower part of the genito-urinary tract, liver and pancreas.

The tissues of animals are usually classified into 5 groups. (1) Epithelial tissue (2) Connective tissue (3) Muscle tissue (4) Nervous tissue (5) Blood and lymph.

The epithelial tissues form sheets to cover organs and lining cavities. Examples are the skin and the lining of the G.I. tract and lungs. They can be classified on the basis of morphology as squamous, cubical, columnar, ciliated and secretory. Malignant growth in these tissues are called carcinoma or more precisely adenocarcinoma, epithelioma etc.

The examples of structural connective tissues are bone, fibrous tissue, cartilage, tendon etc. and characterized by specific intercellular material or matrix. Malignant growths of these tissues are called sarcoma or more specifically fibrosarcoma. Sarcomas are the tumours of skeletal muscle and smooth muscles (intestine and heart muscle).

Brain, spinal cord, peripheral nerves and ganglia constitute nervous tissues. It is highly complex and includes supporting cells, called glial cells e.g. astroglia, microglia and oligodendroglia. Malignant growths of nerve cells called glioma, astrocytoma etc which are rare in adults but common in infants.

The blood and lymph tissues include all the cells in the peripheral blood and their precursors in the bone marrow and lymph glands. Malignant tissues of these tissues are called leukemias.

It is apparent that tissues of ectodermal and endodermal origins have close similarities and malignant growths of these tissue are called carcinomas whereas mesodermal origin have a variety of functions and malignant growth are called sarcomas.

Growth, differentiation and metabolism of cells in culture

There are three different terms namely tissue culture, organ culture and cell culture. In a tissue and organ culture, a small fragments of tissue are placed in medium and allowed to develop. In tissue culture, tissues are allowed to disorganize whereas in organ culture such measure is not taken. In cell culture, the tissues is intentionally disorganized by disrupting it into individual cells by trypsin.

Primary and established cell lines

To obtain a primary cell culture, tissues are cut in small fragments and enzymatically digested into constituent cells. Besides, the specialized cells of the tissue, connective tissue cells, blood cells and reticuloendothelial cells are also present. Most of the blood cells die and disappear from the culture within 2-3 days. Neurons and muscle cells persist in culture for months only without dividing at all. Still other cells begin to divide rapidly and continue to do so for sometime. If the cells multiply repeatedly for a long time they can be passaged. This is usually achieved by first obtaining the cells in suspension and then put into the new plastic bottles with fresh medium. As soon as cells have been passaged in this way the culture is called primary cell line. Primary cell lines go on dividing at a high rate for a long time and can be passaged repeatedly. Even after a large number of passages some primary cell lines cease to proliferate and die out. However, sometimes cell lines have developed the potential to be subcultured indefinitely *in vitro*. Such cell lines are called established cell lines. A line is not designated an established cell line if it has been subcultured atleast 70 times.

The transition from primary cell line to established cell line is smooth and gradual and difficult to say the interface between primary cell line and established cell line. But in other cases, established cell line arises from a primary cell line by a dramatic event called cell alteration and occur in primary cell line which is growing very slowly. Suddenly, a few rapidly growing colonies of apparently altered cells appear to be predominant and outgrow the rest of the culture. Established cell lines arising in this way often differ in many respects from primary cell lines from which they are derived. Primary cell lines maintain many of the characteristics of the cells of their origin whereas established cell lines diverge from these. All the primary cell line have normal number of chromosomes but established cell lines often diverge from these and have an unusual number.

The human fibroblast is an example of primary cell line and never undergoes spontaneous transformation to an established cell line whereas mouse fibroblasts undergo a smooth transition to form established cell lines. The human fibroblasts are diploid, having fibroblastic morphology with a doubling time of 1.5 to 3 days. It shows the phenomenon of contact inhibition, multiply until they reach a maximum density and they can be subcultured for many months. Although human fibroblasts never undergo spontaneous transformation they can be transformed by treatment with certain tumour viruses and colonies of rapidly growing cells outgrow the parent culture. These cells rapidly develop abnormal chromosome number.

Mouse fibroblasts behave in a different way. The multiplication rate of these cells starts declining almost very soon. However, after about 3 months in culture they show evidence of spontaneous transformation and begin to multiply faster. The chromosomes are at first perfectly normal but after a few passages become aneuploid.

Established cell lines irrespective of their origin behave in a similar fashion with a doubling time of 12-20 hours. They are aneuploid and can grow to much higher densities than primary cell line. They have similar nutritional requirement and can grow in suspension culture whereas primary cell lines do not.

Several factors have been identified as the cause of transformation. Both DNA viruses (Polyoma virus) and RNA viruses (Rous sarcoma virus) can cause the transformation of primary cell line. The treatment of cell culture with carcinogenic chemicals viz., 3-4 benzpyrene, 6,4 fluoronitroquinoline –N-oxide, N nitrosomethylurea and 20 methylcholanthrene and X-ray irradiation can cause transformation of primary cell line.

Characteristics of cell lines

(a) *Primary cell line* : Primary cell lines behave quite differently compared to established cell lines. It has been observed that cultures of a primary cell line stop growing before the medium is exhausted which is not true in case of established cell line. Certain cells mainly fibroblasts when come in contact with each other are immediately immobilized and this phenomenon is called contact inhibition. It is mainly a property of primary cell lines and cell division, DNA synthesis, RNA synthesis and protein synthesis are reduced or eliminated when a confluent cell sheet is formed. It is a control mechanism in the intact animal as well.

(b) *Established cell line* : The growth of established cell lines stops due to exhaustion of nutrients or accumulation of toxic products mainly hydrogen ions. One of the outstanding feature of the transformed cell is the absence of great diminution of contact inhibition. That is why transformed cells grow to higher densities and show pile up appearance.

Genetics of cultured cells : Primary cell lines usually retain diploid karyotype whereas transformed cell lines show great variation in karyotype. Shortly after transformation, the incidence of tetraploid cells increases followed by the appearance of aneuploid cells. In the early stages of an established cell line, the numbers of chromosomes may abnormal but the morphology of individual chromosomes is normal. Later, due to breakage, fusions and translocations of chromosomes, the morphology of chromosomes become unrecognizable. For example, many mouse cell lines contain metacentric chromosomes but in mouse all the chromosomes are normally acrocentric.

3
THE CELL AND ITS ENVIRONMENT

The best environment for growing cells is similar to the conditions what they experience *in vivo*. A number of various environmental factors affecting the tissues are :

(1) *Temperature* : In living matter, complicated chemical reactions are happened at comparatively low temperature. In most of the mammalian and avian tissues the temperature is 37°C to 38.5°C. If the temperature is enhanced to 45°C, cells are destroyed within an hour but survive 12-24 hours at 42°C. It is applicable to most of the fibroblasts and epithelial types of cells arising from mammalian tissues. Some of the mammalian cells (human skin epithelial cells) and amphibian cells prefer lower temperature but fish tissue will not survive beyond 20°C.

Most of the cells will survive cooling to a considerable extent and grow slowly at 20-25°C. They can be even stored at 4°C for some time without apparent harm. Although some cell types are sensitive to this cooling but majority are unharmed. If cells are cooled below freezing point, they are destroyed due to formation of ice crystals within the cytoplasm. But if a protective agent such as glycerol is added to the medium, cells are frozen slowly to a very low (-70°C) temperature and may be stored for few months. If the cells are kept in the liquid nitrogen (-196°C) along with cryoprotective agent, cells can be revived even after several years.

(2) *Osmotic pressure* : The osmotic pressure of mammalian cells at 38°C is about 7.6. Cells are not affected by variations of + or – 10%. The osmotic pressure in most biological fluids is mainly due to crystalloids and in animal tissue mainly by NaCl. Other ions particularly glucose also contribute substantially in the change of osmotic pressure. The larger molecule in the medium contribute little to the osmotic pressure.

(3) *H+ concentration* : Although neutral pH of biological fluid is essential for the survival of the whole animal, the isolated tissues tolerate remarkably wide variations (pH 6.8 to 7.8). Some cells growing at pH above 8.0 or pH below 6.6 have been reported. But optimal growth is obtained between pH 7.2 to 7.4 and it is not desirable to deviate the limits of pH 6.8 to 7.8. Most of the cells die at pH above 7.8 or below 6.8.

Other inorganic ions : In case of most animal cells the ions required are sodium, potassium, calcium, magnesium, iron, carbonate, phosphate and probably sulphate. The sodium and potassium ions are essential for the maintenance of osmotic pressure in the medium. Calcium and magnesium ions are required for the function of certain intracellular enzymes. Both calcium and magnesium are also necessary for the binding of the cells to the glass or plastic surface. Iron is required for cytochrome. The bicarbonate ion besides having the buffering property is necessary for many biochemical processes. The phosphate ion is essential for energy metabolism and buffering function.

Carbohydrates : The principal source of carbohydrates is glucose although it can be replaced by fructose, mannose, maltose, trehalose, glucose phosphates and possibly galactose. Disaccharides and polysaccharides are first broken down into monosaccharides before being utilized. Glucose can be replaced by lactate and pyruvate as energy source as long as plenty of O_2 available. Some mutant strains can utilize pentoses such as xylose and ribose. Some cells can survive without glucose because amino acid residues may be used as source of energy after deamination.

Gases : Both O_2 and CO_2 are essential for cell survival. However, it seems unlikely that any animal cell will survive indefinitely in the absence of O_2 since some of the by products of Krebs cycle are necessary for the synthesis of various cell components and unless these are supplied in the medium the cells are bound to die when the stores are exhausted. In some cell types the bicarbonate ion is absolutely essential. Apart from its requirement in tissue metabolism the bicarbonate ion is the most important buffering ion in most culture media.

Amino acids : Most of the animal cells have a specific requirement of 12 amino acids necessary for cell growth and multiplication. These are arginine, cystine, histidine, isoleucine, lysine, methionine, phenylalanine, threonine, tryptophan, tyrosine and valine. Besides, most cells have a high requirement of glutamine necessary for synthesis of nucleic acids.

Vitamins : Several vitamins of B groups viz., paraaminobenzoic acid, biotin, choline, folic acid, nicotinic acid, pantothenic acid, pyridoxal, riboflavin, thiamine and inositol are necessary for cell growth and multiplication. Nicotinic acid can be replaced by nicotinamide and pyridoxal can be replaced by pyridoxine. Other type of vitamins do not appear to be essential for cell survival.

Proteins and peptides : It is not established to what extent proteins and peptides are essential for growth and survival of cells. Media manufacturing companies have prepared few media in which cells will grow rapidly in the absence of protein or similar polypeptides. In the few cases, growth in protein free media has been achieved by the addition of a trace of serum.

Buffering system : In inorganic bicarbonate CO_2 buffer system, cultures may become alkaline very quickly after removing the culture from the incubator. The organic buffer system used to maintain culture pH are HEPES (pKa = 7.3 at 37°C) and MOPS (pKa =7.0 at 37°C) at a concentration of 10-20 mM without an enriched CO_2 atmosphere. Using HEPES, the CO_2 level can be reduced to around 2% with concominant decrease in bicarbonate concentration.

4
TYPES OF CELLS AND CELL CULTURE

Types of cells : Animal cells are usually defined by the tissue of their origin. On the basis of morphology, they can be divided into 5 types.

1. *Epithelial cells* : Epithelial tissues consist of a layer of cells which cover organs and line cavity. Example skin and the lining of alimentary canal. They grow well as single cell monolayer in culture.

2. *Fibroblast cells* : Connective tissue form the major structural component of animals. The tissue contains fibroblasts and most widely used cell in the laboratory. Fibroblasts are bound to the fibrous protein collagen in the connective tissue. They have excellent growth characteristics.

3. *Muscle cells* : Muscle tissue consists of a series of tubules formed from precursor cells which fuse to form a multinucleate complex and contain structural protein actin or myosin. The precursor cells are myoblasts which

are capable of differentiation to form myotubes – a process observed in culture.

4. *Nerve cell* : Nervous tissue consists of neurons and glial cells. Neurons are responsible for the transmission of electrical impulses. Neurons are highly differentiated and do not divide in culture. However, the addition of nerve growth factor to cultures of neurones may cause the formation of cytoplasmic outgrowths called neurites. Neuroblastomas are tumor cells and undergo growth in culture.

5. *Blood and lymph* : They are a range of cells in suspension.

Characteristics of a normal cell

1. *Diploid chromosome number* : Normal cell has a diploid chromosome number (46 in human cells). It means there is no gross chromosomal changes.

2. *Anchorage dependence* : The cells require a solid substrate for attachment and growth. In the laboratory, petridish, tissue culture flask, tissue culture plate, MD bottle and Roux flasks are usually used as substrate. There is electrostatic attraction and Van der Waal's forces between cell membrane and substrate. Cell adhesion occurs by divalent cations (Ca^{++} and Mg^{++}) and basic protein forms a layer between the solid substrate and cell surface. Serum derived glycoprotein (fibronectin) can provide a surface coating conducive to cell attachment. Conditioning factors are released by cells into the medium and help in forming a bond between cell surface glycoprotein and substrate. The density of the electrostatic charge on the solid substrate is also critical in improving cell attachment and negatively charge is provided on glass surface by alkali treatment. Tissue culture grade plasticware consists of sulfonated polystyrene with 2-5 negatively charged groups per nm^2 surface area. The cells

grow until a confluent monolayer of cells is formed on the substrate.

3. *Non-malignant* : The cells are non-neoplastic. They do not produce any tumour following injection into immunocompromised mice.

4. *Finite lifespan* : It has a finite life span.

Characteristics of a transformed cell : It means change of animal cells from normal to infinite growth capacity. Normal animal cells have a growth capacity but some cells acquire a capacity for infinite growth and such a population can be called an established or continuous cell line. This requires cells to undergo a process called transformation and lose the sensitivity to the stimuli associated with growth control. Transformed cells may also lose their anchorage dependence and often show changes in number of chromosome (aneuploidy). The transformed cells have a high capacity to grow in relatively simple growth medium and even without growth factors.

Carcinogenesis *in vivo* is analogous but not identical to transformation of cells *in vitro*. Not all transformed cells are malignant characterized by ability to form tumours in animals. On the other hand all tumour derived cells grow continuously in culture. Hela and Namalwa are examples of human cervical cancers and human lymphoma cases respectively. Cells can be transformed or immortalized by treating with mutagens, viruses or oncogenes.

Types of cell culture : Many types of cells undergo only a few divisions *in vitro* before dying out, whereas others will survive for up to a hundred cell generations and some can be propagated indefinitely. These differences, the nature of which are not fully understood, give us four main types of cultured cells.

a) *Primary cell culture* : When cultures are established initially from tissue taken directly from animals, they contain several cell types, most of which are capable of only 5-10 divisions. Due to high cost, inconvenience of getting fresh tissue each time and variation from batch to batch, it is not suitable for use in routine diagnostic work or vaccine production. Furthermore, the donor animals may harbour the latent viruses which can confuse diagnosis or contaminate vaccines. But the primary cultures are very sensitive to many human and veterinary viruses and still used for primary isolation of these viruses. Primary cell cultures retain some of the characteristics of the tissue from which they were derived and usually contain more than one cell type. With a few exceptions (such as nerve and muscle cells) cells in culture can be classified in two general morphological types. Fibroblast like cells are thin and elongated, while epithelial like cells are polygonal in shape and tend to form sheets. Due to the limited life span of primary and secondary cell cultures, one can not do long term experiments with these.

b) *Secondary cell culture* : When primary cells are again passaged *in vitro*, it is called secondary cell culture.

c) *Diploid cell strains* : These are cells that are capable of undergoing a number of divisions in culture that is routinely related to the life span of the species of animals- about 50 for fetal human cells and about 10 for fetal cells from horses and cows. Diploid cell strains of fibroblasts established from human fetuses or embryos are widely used in human diagnostic virology and vaccine production, but diploid strains have not been much used in veterinary vaccine production. Upon serial transfers of primary cells, a gradual selection may occur until a particular cell type become predominant. If these cells continue to grow at a constant rate over a successive passages, these primary cells are referred to as a cell strain. These cells have a finite life span of less than 100 division *in vitro*. Diploid cell strains that retain their original diploid chromosome number and that are non-malignant are useful in vaccine production.

d) *Continuous cell line* : These are cells of a single type that are capable of indefinite propagation *in vitro*. Such immortal cell lines originate from cancers or by spontaneous transformation of a diploid cell strain. They do not bear the close resemblance to their cell of origin as they undergo many mutations during their prolonged culture. The cells usually have lost the specialized morphology and biochemical abilities. Cells of continuous cell lines are often aneuploid in chromosome number, especially if of malignant origin. Continuous cell lines derived from monkey (Vero), dog (MDCK), cattle (MDBK), pig (PK15), Cat (CRFK), mouse (L929, 3T3), hamster (BHK21), rabbit (RK-13), and others are widely used in experimental and diagnostic virology. The great advantage of continuous cell lines over primary cell cultures is that they can be propagated indefinitely by subculturing the cells at regular intervals. Like other cells, they can be preserved for many years when frozen in serum containing medium with added glycerol or DMSO and stored at -80^0C or -196^0C (liquid nitrogen). If the cells in cell strain undergo a transformation process (spontaneous or induced changes in karyotype, morphology or growth properties), that makes them immortal (unable to divide indefinitely) they are called a cell line. A cell line is derived from a cell strain by infection with oncogenic viruses or by exposure to chemical carcinogens. Cell lines often have abnormal chromosome numbers and may be tumorigenic when inoculated into susceptible animals. Cell lines do not exhibit property of contact inhibition (inhibition of growth under crowded condition) but continue to show pile up appearance.

Each culture system possesses advantage not shared by the other so that a choice may be possible to suit particular circumstances when the cells along with media are to grow in glass or plastic bottle, after a period of time cells attach to the bottom of the flask and start diving until a monolayer (one cell thick continuous layer) is formed. Cells that require a surface for attachment are referred to as anchorage dependent whereas certain cells (lymphoid cells) are not anchorage dependent and are cultivated in suspension culture.

Source of cells : Well known international and national animal cell culture collections are :

International
1. The American Type culture collection (ATCC), 12301 ParKlawn Drive, Rockville, Maryland 20852, USA.
2. The European collection of Animal cell culture (ECACC), Public Health Laboratory Service (PHLS), Centre for Applied Microbiology Research (CAMR) Porton Down, Salisbury, SP4 0JG, U.K.

National
1. Cell Repository – National Centre for Cell Science, NCCS Complex, Univeristy of Pune Campus, Ganesh Khind, Pune – 411 007.

5

CELL CULTURE : APPLICATIONS ADVANTAGES & DISADVANTAGES

A finite growth capacity is a characteristic of all cells derived form the normal animal tissue. The cells pass through a series of age related changes before being incapable to divide further. The finite number of generations of growth is a characteristic of the cell type, age and species of origin and referred to as the "Hayflick limit". The cells derived from embryonic tissue have a greater growth capacity than those derived from adult tissue.

Applications of animal cell culture

There are a number of applications of animal cell culture. Cell cultures provide a good model system for studying (1) basic cell biology and biochemistry (2) the interactions between disease-causing agents and cells (3) the effects of drugs on cells (4) the process and triggers for aging and (5) nutritional studies besides a number of other applications given below :

Toxicity testing

Cultured cells are widely used alone or in conjunction with animal tests to study the effects of new drugs, cosmetics and chemicals on survival and growth in a wide variety of cell types.

Cancer research

Since both normal cells and cancer cells can be grown in culture, the basic differences between them can be closely studied. In addition, it is possible, by the use of chemicals, viruses and radiation, to convert normal cultured cells to cancer causing cells. Thus, the mechanisms that cause the cancer can be explored. Cultured cancer cells also serve as a test system to choose the suitable drugs and methods for selectively destroying types of cancer.

Virology

One of the earliest and major uses of cell culture is the replication of viruses in cell cultures (in place of animals) for titration, SNT and in vaccine production. Cell cultures are also widely used in the clinical detection and isolation of viruses, as well as to know the basic mechanisms between cell and virus interactions.

Cell-based manufacturing

Cultured cells can be used to produce many important products. The large-scale production of viruses for use in vaccine production is the most important one. It is also used for large-scale production of cells that have been genetically engineered to produce proteins that have medicinal or commercial value. These include monoclonal antibodies, insulin, hormones, etc. Again, the use of cells as replacement tissues and organs can be performed. Artificial skin for use in treating burns and ulcers is the first commercially available product. The research is going on to make artificial organs such as pancreas, liver and kidney. A potential supply of replacement cells and tissues may come out of work currently being done with both embryonic and adult stem cells. These are cells that have the potential to differentiate into a variety of different cell types.

Genetic counseling

The collection of amniotic fluid from the pregnant women can be very useful to diagnose any abnormality in the foetus. These cells can then be examined for abnormalities in their chromosomes and genes using karyotyping, and other molecular techniques.

Genetic engineering

The ability to transfect the cultured cells with new genetic material (DNA and genes) has provided a major tool to express theses genes (new proteins). These techniques can also be used to produce these new proteins in large quantity in cultured cells for further study. Insect cells are widely used as cell factories to express substantial quantities of proteins that they manufacture after being infected with genetically engineered baculoviruses.

Gene therapy

The ability to genetically engineer cells has also led to their use for gene therapy. Cells can be removed from a patient lacking a functional gene and the missing or damaged gene can then be replaced. The cells can be grown for a while in culture and then replaced into the patient. An alternative approach is to place the missing gene into a viral vector followed by inoculation into the patient and the missing gene will then be expressed in the patient's cells.

Drug screening and development

Cell-based assays have become increasingly important for the pharmaceutical industry, not just for cytotoxicity testing but also to screen a large number of compounds that have potential use as drugs. Originally, these cell culture tests were done in 96 well plates but rapidly replaced by 384 and 1536 well plates. So, in brief the applications of the cell culture are :

1. It is used to investigate the normal physiology or biochemistry of the cells.
2. It is used to evaluate the effect of compounds (metabolites, toxins, cadmium, arsenic, hormones, growth factors, mitogens etc.) on the cell culture.
3. It is used to produce artificial tissue (skin for burn patients).
4. It is used to produce the large quantity of viruses by propagating in cell culture for biochemical and molecular characterization.
5. It is used to carry out the virus induced plaque assay, pock assay etc.
6. It is used for virus neutralization test and virus titration.
7. It is used to synthesize valuable biologicals (hormone, recombinant protein, growth factors) in large quantity for diagnostic, therapeutic and other uses.
8. It is used for propagation and large scale production of viral vaccines.
9. It is used for large scale production of protein through transfection for diagnostic, prophylactic and therapeutic purposes.
10. It is used for production of monoclonal antibodies for diagnostic, therapeutic and other uses.
11. It is used for production of recombinant glycoprotein of pharmaceutical, medical and veterinary importance.

Advantages of cell culture

1. Most consistent and reproducible results can be made available by using same batch of cell and homogenous population of cells.

Cell Culture : Applications, Advantages & Disadvantages

2. The effect of a toxin on a particular organ (eg liver) can be evaluated more accurately in the liver cell culture than in the laboratory animals.

3. It is less expensive than whole animal.

4. The production of a particular biological can be obtained in more pure form in cell culture than from pooled organs of animals.

5. The requirement of manpower, time and money would be less in cell culture than in whole animal experiments.

6. The virus titration can be easily and reliably carried out in cell culture compared to whole animals.

7. The pock and plaque morphology and characteristics can be easily carried out in cell culture.

8. Virus neutralization test can be easily carried out in cell culture than whole animals and will be less expensive.

9. The cytopathic nature of the virus can be evaluated in the cell culture.

Disadvantages

1. The skill, knowledge and expertise of the persons should be up to the mark to maintain the cell culture in the laboratory in perfect condition.

2. Cell culture laboratory should be highly sophisticated and desirably having dust proof and hepa filtered air circulation facility.

3. The contamination of cell culture with bacteria, fungi and mycoplasma is a major problem.

4. After a period of continuous growth, cell characteristics can change and may be quite different from original one.

5. Cells can adapt to different nutrients leading to changes in intracellular enzymes.

6. Fast growing cells often overgrow over the slow growing cells in mixed cell population.

7. The activity of intracellular enzyme activities change dramatically in the event of nutrient depletion and by product accumulation in a culture.

6
IDEAL CELL CULTURE LABORATORY WITH EQUIPMENTS

A cell culture laboratory should allow the sterile handling of cultured cells with no or minimum level of contamination. The growth of bacteria and fungi is very high and no level of contamination can be allowed in animal cell culture. However, it is impractical to design a laboratory free from potentially contaminated microorganism. Further, it is possible to establish a laboratory with minimal risk of contamination. Several points should be taken into consideration while designing a new cell culture laboratory.

(1) There should be a good physical separation of sterile handling area from area of washing and sterilization.

(2) The cell culture facility are often located in small rooms with minimal traffic.

(3) A class II laminar flow is ideal as air flow through the sterile exhaust of the cabinet should maintain a low particle count in the environment.

(4) The essential equipments such as microscope and incubators should be kept close to laminar hoods to minimize physical transfer of the culture.

(5) The incoming air into the culture room should be filtered through a high efficiency particulate air (HEPA) filter. The incoming air pressure may increased to cause a slight positive pressure within the laboratory.

(6) Air conditioning may be necessary particularly during summer season.

(7) Doorways should be wide enough and ceiling should be high enough to provide a clearance for installation of instruments such as laminar hood, incubators, autoclave etc.

(8) Washing and sterilization facilities should be in a different room outside the culture laboratory.

In an ideal cell culture laboratory, room should be of adequate size to accommodate the necessary equipments and isolated from other rooms preferably by one/two doors. The space between these two doors should be used as changing room. In the changing room, normal dresses are changed with laboratory coats used only in the tissue culture room. The windows in the doors allow observation of personnel within and outside the laboratory. This can help prevent both doors opening simultaneously and minimizes the entry of possible contaminants into the laboratory. The atmospheric pressure in the tissue culture laboratory should be at a higher pressure (positive) than the outside to prevent inflow of air. The floor should be made up of non-porous light coloured tiles/blocks which are easy to clean. The laboratory should have glass windows for better visibility and easy examination of cell suspension. Air-conditioning system is necessary in the cell culture laboratory and should be properly placed so that there is no interference of the airflow between it and hood.

Ideal Cell Culture Laboratory with Equipments

The design of a virology laboratory with cell culture facility depends on the purpose and requirement for which the laboratory is beings established.

Measurements of a ideal cell culture laboratory

Size of laboratory unit	24-30 m^2
Size of the booth	2.5X2 m^2
Size of autopsy room	50X80 m^2
Ceiling height	3 m
Door height	2.25 m
Door width	1 m
Bench height	0.7-.75 m
Bench width	0.6-0.8 m
Electricity	220-240 V, Ac 60 H3
Illumination of working are	260-240 lumen/ m^2
Volume of fresh air/person/hours	60 m^3
Ideal room temperature	22-26°C
Ideal room humidity	50-60%
Small animal room temperature	23-26°C

The equipments needed in a cell culture laboratory are :
Equipments
A. Tissue culture hoods

Two points are considered in designing the hoods (i) protecting the tissue culture from the operator (ii) protecting the operator from the tissue culture. Different types of hoods available are : (a) Laminar flow (b) Class I (c) Class II and (d) Class III.

(a) Laminar flow hoods do not provide any protection to the operators/tissue culture worker. The cabinet is open and a stream of filtered and sterile air flows from the back over the workplace towards the operator. Obviously, any cell culture with potential risk or health hazard (potentially infectious organisms, radioactive or toxic or volatile solutions) should never be used in these type of hoods. It is mostly used in filtration of cell culture media.

(b) Class I hoods provide good protection to the operator and to a lesser degree to the cell culture. Air is drawn from the open front over the cell culture and out through the top of the hood. These hoods are used in sterile work areas/rooms. The Class I Biosafety cabinet is a ventilated cabinet with an inward airflow and outlet HEPA filters. It was previously referred to as the CDC Hood and served a valuable function in its time by protecting personnel and the environment. Because it offers no product protection, it has been essentially obsolete for the past several decades.

(c) Class II hoods offer protection to the operator and cell culture. Filtered air is drawn in through the top of the hood down over the tissue culture and through the bottom of the work area. Additionally, air is drawn from the half open front past the operator and down through the grill in front of the work area. The class II hood is the most common type found in a tissue culture laboratory. Class II (types A and B) are Laminar Flow Biological Safety Cabinets that protect personnel, product and environment. They provide inward airflow to protect personnel, downflow HEPA filtered air to the work area to protect the product. A class II laminar hood is suitable for work with low to moderately toxic or hazardous infectious agents. There is an inward flow of air drawn into the sterile working area through HEPA filter. The air curtain formed in the internal front face of the cabinet has a typical flow rate of 0.4 m/sec. The exhausted air is also forced through HEPA filter and serves to protect the surrounding environment from any potentially pathogens or toxic compounds. The HEPA filters in a class II cabinet ensures a 99.9% efficiency of entrapment of 0.3 μM particles.

(d) Class III hoods are totally enclosed system found in specialized laboratory and are used to handle the highly pathogenic organisms. In this type of hoods, the operator is protected from the highly pathogenic organisms by a full physical barrier as the open front is blocked by glass or Perspex and a pair of heavy duty protective gloves attached through which the work is accessed. The exhausted air from the cabinet is filtered through at least two HEPA filters to ensure the complete removal of all pathogens. All equipments entering the cabinet is passed through an air lock and removed directly into an autoclave. The Class III cabinet is defined as ventilated Glovebox. This is a gas-tight chamber operated through sealed gloves which provides a complete barrier between the worker and hazardous material. The Glovebox is maintained under negative pressure with HEPA filtered supply air and double HEPA filtered exhaust air.

Most laminar hoods are available with UV light for sterilization of the work space when not in use. It is effective with a new UV tube and over the time the effectiveness is greatly diminished. The hoods should be cleaned on a regular basis with 70% ethanol before the start of the work and end of the work.

B. Autoclaves and Hot air ovens

They are not placed in the cell culture room but adjacent room. In autoclaves, steam is commonly used for sterilization. The autoclave is run at 121°C and 15 lb/inch^2 for 15 min. After this period the steam supply is terminated the pressure drops and the autoclave cools. Bacteria, fungi, their spores, viruses, mycoplasma will not survive these conditions as the wet heat is effective method for denaturing proteins. Plastic micropipette tips, water, buffers, polypropylene centrifuge tubes, eppendorf tubes, PCR tubes etc should be sterilized by autoclave. Some liquids such as growth media, glutamine, trypsin can not be autoclaved as their components can be destroyed or denatured under extreme conditions. Plastic items made from polystyrene (most of the plasticware used for cell culture) can not withstand autoclave conditions. Practically, those item that can survive hot air oven conditions can also be put in the autoclave. Bottles to be autoclaved should have their caps slightly loosened for effective sterilization. Metal pipette cans can be sterilized both by dry heat and wet heat. The autoclave tape is used to monitor whether the items have been autoclaved or not. It is initially plain but on exposure to high temperature produces dark stripes across the tape.

In the hot air ovens, the temperatures is set at 180°C and maintained for 1 hour for sterilization. Heat killing (hot air oven) is less efficient than sterilization by wet heat in autoclave. Hot air ovens are useful for some glasswares e.g. pipettes, beakers, test tubes, glass syringes, glass centrifuge tubes, conical flasks of different sizes etc.

C. Incubators

For optimal growth, most animal cell types require a temperature of 37°C in medium at pH 7.0-7.2. A slight decrease in temperature from the optimal may slow the cell growth rate but an increase in temperature is more detrimental to the cells. Cells will survive at 39°C for few hours and they will die rapidly at temperature above 40°C. In inorganic bicarbonate-CO_2 buffer system, cultures may become alkaline very quickly after removing the culture from the incubator. The maintenance of a humidified atmosphere is essential to prevent loss of medium in non-sealed culture system such as petridish, tissue culture plate and flask without airtight cap. In the CO_2 incubator, there is a constant supply of CO_2 usually at 5% level. The low level of CO_2 causes an increase in culture pH whereas too high a level of CO_2 causes an decrease in pH. To counter the acidic pH due to production of by products of cellular metabolism, a buffering system is used. Most commonly used system is bicarbonate/CO_2 system requiring supply of CO_2 to the incubator and bicarbonate in the medium. However, some laboratories maintain dry nongassed incubators using alternative buffering system namely morpholinopropane sulfonic acid (MOPS) or N-2 hydroxyethylpiperazine -N'-2 ethanesulfonic acid (HEPES). CO_2 incubators control both the internal temperature (37°C) and the gas mixture in the chamber (5% CO_2 +95% air). The 5% CO_2 must be carefully controlled as it governs the stability of the buffering system. The bicarbonate added to the medium (sodium hydrogen carbonate) effectively mops up the acidic ions forming carbonic acid which is in equilibrium with the water and dissolved CO_2 to maintain a pH of 7.0-7.2. In dry non-gassed incubators internal temperature is set at 37°C. The culture bottles are grown in closed system i.e. caps of the culture flasks are tightened. However, it can not be used to culture cells in petridishes, tissue culture plates as evaporation of the medium can harm the culture. However, they can be put in the closed incubators inside the desiccators.

The organic buffer system HEPES has a pKa value of 7.3 at 37°C and MOPS has a pKa value of 7.0 at 37°C and can be used at a concentration of 10-20 mM without an enriched CO_2 atmosphere. Using HEPES, the CO_2 level can be reduced to around 2% with concomitant decrease in bicarbonate concentration. Both inorganic or organic buffer system with a high degree of pH control can be used. But organic buffers are expensive and not routinely used. The CO_2 incubator has become an indispensable standard piece of item in a cell culture laboratory. The CO_2 level is maintained at 5%. The incubators and the laminar hoods should be cleaned with 70% ethanol.

D. Centrifuges

For most cell culture, only low speed centrifuges are required. In most of the cases cells are centrifuged at 150-200g for 5-10 min at 20°C. However, refrigerated centrifuges can be used to avoid exposing cells to uncontrolled higher temperature due to generation of heat during centrifugation. The performance of the centrifuge is expressed as the relative centrifugal force (RCF) or g and is calculated from the following formula.

RCF (g) = 1.118 X 10^{-5} X R X N^2

R = distance in cm from the centre of centrifuge shaft to extended tip of the centrifuge tube

N = revolutions of centrifuge per min (rpm)

For example rpm = $\sqrt{\frac{g \times 10^8}{R \times 1118}}$, if r = 25 and g = 500

rpm = $\sqrt{\frac{500 \times 10^8}{25 \times 1118}}$ = 1337

E. Microscope

It is preferable to have both an inverted microscope and a standard microscope within the cell culture laboratory. An inverted microscope is valuable for visualizing cell cultures in solution and to know their growth and health. Both the microscope with a movable slide holder is needed to move the tissue culture plates and counting chambers.

F. Water and media filtration devices

For all cell culture reagents, HPLC grade water is recommended. A number of filtration systems that produce HPLC grade water are available commercially. Triple glass distilled water can also serve the purpose. The water purified earlier should not be used as the minimal microbial growth can lead to pyrogen contamination. Algae can grow anywhere and microorganisms that get benefits from the algae can develop quickly. For large scale filtration, pump driven or pressure driven (positive or negative) devices are available. For small scale filtration sterile plastic vacuum filtration devices both disposable and non-disposable are available.

G. Liquid nitrogen storage tanks

Cells can be stored in liquid nitrogen for several years. It is necessary as all the time animals are not available for primary culture. Further, preserved cells can be of great help in case of contamination of cells by bacteria, fungi or mycoplasma. For a valuable cell line – a master bank and working cell bank are maintained. In the master cell bank, cells at low passages are maintained whereas in the working cell bank cells formed by growth of several passages of master cell are maintained. Working cell bank is commonly used whereas the master cell bank is accessed only when it is absolutely necessary. Cells can be stored at a concentration of 10^7 cells in a freezing medium containing 7 ml EMEM + 2 ml FCS + 1 ml of DMSO or glycerol. DMSO or glycerol will protect the cells from disruption during freezing and thawing process. The cells can be stored indefinitely in liquid nitrogen at -196°C if the level of nitrogen is maintained properly. To maintain the high viability of the stored cells, freezing and thawing should be properly followed. Slow freezing and rapid thawing is generally recommended for maximum cell survival. The cells are stored in 1ml/2ml cryovial. The cell suspension can be frozen by placing the ampoules containing the cells in a polystyrene box at -70°C overnight. This ensures an initial freezing rate of about 1°C/min followed by transfer of ampoules into the liquid nitrogen. Programmeable coolers are also available to control the rate of cooling. Liquid nitrogen cans are available in the capacity of 25-500 litres and may be narrow or wide necked. The cryovials are placed in large drawers in wide necked freezers or to metal canisters in narrow necked freezers. Cryogenic plastic vials have strong seals to prevent leakage that could result from large temperature fluctuation. The level of liquid nitrogen in the storage container should be monitored and maintained properly. Liquid nitrogen tanks for storage of cells vary from relatively small to very large with or without an automatic nitrogen level sensing and filling capability. It is easier to retrieve cells from small tanks. Cells can be stored both in liquid and vapour phase in the liquid nitrogen container. If the cells are stored in vapor phase nitrogen, vials are less likely to fill with liquids, explode during thawing and be contaminated by microorganisms via liquid.

H. Water baths

Routine ordinary water baths are generally used in the cell culture laboratory. Water baths have temperature controlling devices and can be set over a wide ranges of temperatures. Water baths are a major source of contamination in a cell culture laboratory and should be periodically cleaned and antimicrobial detergents added. Water baths are available in different sizes and with or without water circulation facilities.

I. Plastic wares

Previously, culture flasks made up of borosilicate glassware were used in cell culture (Roux flask, MD bottle, test tubes). However, in recent days, presterilized plastic flasks suitable for cell culture are commercially available and used by most of the laboratory to minimize the contamination. The plastic polystyrene is treated to produce a surface suitable for cell attachment and growth. They are sterilized by γ irradiation and suitable for single use. The attachment of negatively charged cell surface to a negatively charged cell surface require the help of divalent cation such as Ca^{+2} and Mg^{+2} in the medium. Electron microscopic study showed the presence of a 50 $A^°$ protein layer secreted by surface active protein or provided by the serum between cells and the substrate.

The polystyrene used in making cell culture bottles are unsuitable for cell attachment because of hydrophobic nature and without any charge. The plasticwares are treated to produce hydrophilic, wetable and negative surface charge. It is done with either sulfuric acid to allow sulfonation or high voltage electric arc usually density of $2\text{-}10 \times 10^{14}/cm^2$. They can also be modified by applying a variety of anionic and cationic groups (carboxyl and amine groups). The plastic bottles can also be coated with various positively charged polymers such

Ideal Cell Culture Laboratory with Equipments 37

Plastic disposable culture bottles (Size 25 cm^2 to 175 cm^2 surface area)

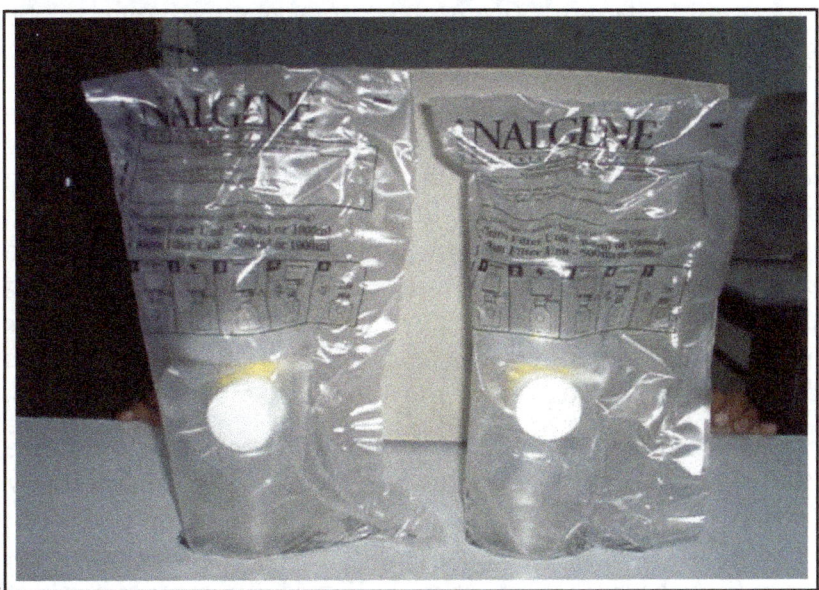

Disposable filtration unit

as DEAE dextran, polyacrylamide, polylysine, polyornithine, polyarginine, polyhistidine and protamine. Poly-D-lysie with molecular weight of 30-300 KDa has been extensively used to coat the glass or plastic bottles/flasks. A solution of polylysine (10 µg/ml) is generally used to wash the bottle sufficient to apply the polymer coating. Most of the vessels offer plastic surface for cell attachment and may be grown in suspension or surface attached cells. The most commonly used plastic culture containers are multiwell plates that can accommodate many replicates of small volume cultures. The 24 well plates hold 3 ml/well and are suitable for cell growth, toxicity or cell stimulation testing experiment. The 96 well plates hold a volume of 0.3 ml per well and suitable for SNT, virus titration and cloning experiment. The plastic culture bottles are available of various sizes eg T-25, T-75, T-175, T-300 etc to have a growth area of 25-300 cm^2. The caps of the bottles are tightly fit and kept in the closed incubators. On the other hands, the caps are slightly loosened and kept in the CO_2 incubator. Some T flasks have five holes with filters that allow gas exchange in the incubators.

The presterilized tissue culture grade plasticware has made the cell culture work easier than using reusable glasswares. It is difficult and time consuming to wash glasswares sufficiently so that it is free of potentially cytotoxic detergent to obtain reproducibility in cell culture. The advantages of single use plastic ware are less preparation time, lower contamination problem and consistency in operation. It is necessary to soak all the reusable glassware in hypochlorite solution as soon as possible after use to prevent the adherence of protein residues.

Osmometer : An important parameter of cell culture medium is the osmotic pressure expressed as osmolarity (number of particles/litres). The effect of each component in the medium to the overall osmolarity is additive and dependent upon its dissociation. For example 15 mM glucose will increase osmolarity by 15 mOsm/litre whereas 15 mM NaCl will increase osmolarity by 30 mOsm/litre. The osmolarity of standard culture medium is approximately 300 mOsm/litre and optimal for most cell lines and a deviation of 10% is tolerable. The osmolarity of a culture may increase during cell growth as a result of the production of low molecular weight metabolites such as ammonia and lactic acid. The osmolarity of the culture can be measured by simple bench top osmometer and based on the measurement of the freezing point of the liquid. Osmolarity measures the freezing point of a solution in comparison to that of water. In a microosmometer, a sample size of 20-50 µl is needed whereas 0.2-2 ml is needed in standard equipment.

7

GROWTH AND MAINTENANCE OF CELLS IN CULTURE

Once cells are obtained from a culture collection or isolated from animal tissue,, cells are put in the flask with growth medium. The cells are inoculated at a density of 10^4-10^5 cells/ml to get a concentration of 10^6 cells/ml or 10^5 cells/cm^2 in a solid surface in 2-3 days because of nutrient limitation, accumulation of toxic metabolite or lacking of growth surface.

Subculture : When cells stop growing in culture, new cultures can be initiated by inoculating some of the cells into fresh medium. This is called subculturing. The cells from one flask are usually split into 3 to 5 flasks. For subculturing of anchorage dependent cells, detachment of cells by treatment with trypsin versene is generally done and placed into new flasks containing fresh medium. Besides, trypsin, collagenase, elastase, hyaluronidase, pronase, dispase either alone or in various combinations can be used. Crude preparations are often more successful than purified enzyme preparations as the

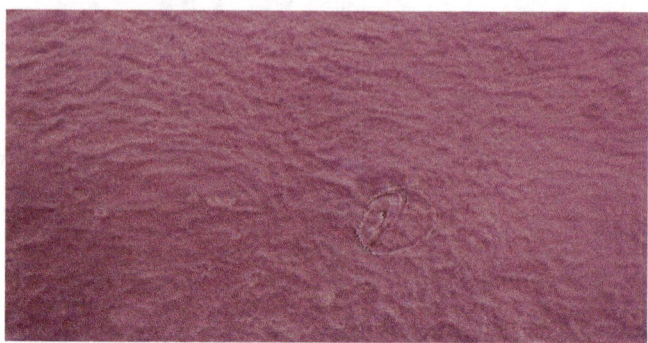
Normal Vero cell line

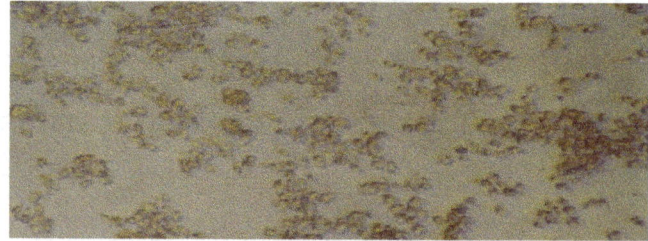
Vero cell line infected with Bluetongue virus

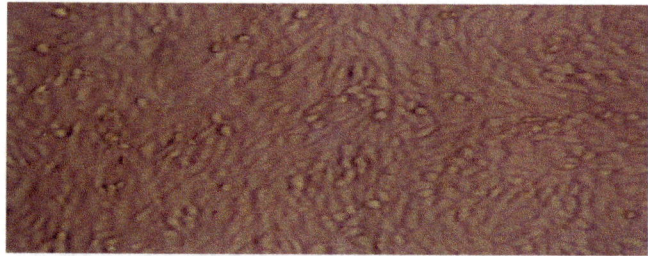
Normal MDBK cell

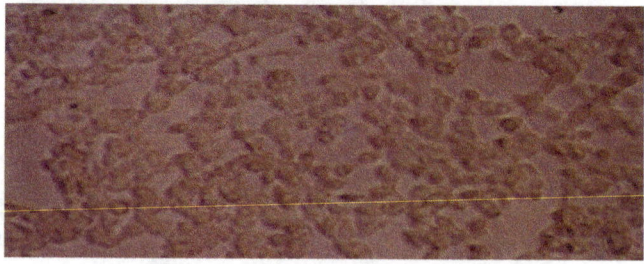
MDBK cell infected with BHV-1

former contain other proteases as contaminants although the latter is generally less toxic and more specific in their action. Trypsin and pronase give the most complete disaggregation but may damage the cells. Collagenase and dispase on the other hand give incomplete disaggregation but are less harmful. Hyaluronidase can be used in conjunction with collagenase to digest the intercellular matrix. Trypsin will break down the proteins that bind the cells to the culture surface. In a 25 cm2 flask 1 ml trypsin is added and kept for 1-2 min at 37°C. Trypsinization is efficient if Ca++ and Mg++ are removed from the medium. This is achieved by preparing EDTA (versene) and trypsin solution in Ca and Mg free salt solution. The over exposure of cells in trypsin is detrimental. The action of trypsin is stopped by the addition of serum supplement medium or centrifugation to remove the cells. If the cells are grown in serum free medium a soybean trypsin inhibitor can be used. Cultures are given a passage number which indicates the number of subcultures performed since the cells are obtained.

Phases of culture : The cell culture follows a growth pattern and can be divided into several phases.

(1) *Lag phase* : This is an early phase and there is no increase in cell concentration. In this phase, there is synthesis of growth factors from the cell. The length of this phase is dependent on the culture medium compositions and initial concentration and state of the cells. The length of lag phase is longer at low inoculation densities of cells with low viability. In case of high density of cells with good viability, the lag phase may be shorter. Transformed cells have a lower requirement for growth factors and often show no lag phase even at low concentration of cells.

(2) *Growth phase* : During growth phase, there is exponential increase in cell number which can be represented by the following formula.

$$N = No.2^X$$

$$\text{or } \log_{10}N = \log_{10}No + X.\log_{10}2$$

N = Final concentration of cells
No = Initial concentration of cells
X = Number of generation for a inoculum of 10^4 cells/ml which reaches to 10^5 cells/ml can be calculated as

$$\log_{10}10^5 = \log_{10}10^4 \; X. \; \log_{10}2$$

$$\text{or } X = \frac{(\log_{10}10^5 - \log_{10}10^4)}{\log_{10}2} = \frac{1}{\log_{10}2} = 3.32 \text{ generation.}$$

The doubling time during cell growth can be calculated from the equation

$$tD = T/X$$

tD = doubling time ,
T = Total elapsed time and
X = number of generations.

Thus for a culture which reaches from 10^4 cells /ml to 10^5 cells /ml in 3 days the average doubling time is 0.904 days or 22 hours.

tD = 3/3.32 = 0.904 days or 22 hours. (3.32 is from above equation)

(3) *Stationary phase* : In this phase there is no further increase of cells concentration and death rate = growth rate. It occurs due to (1) depletion of nutrients in the medium (2) accumulation of metabolic byproducts which is inhibitory to cell growth (3) formation of complete monolayer of the cells over the substrate.

(4) *Decline phase* : This phase occurs due to cell death. There is large difference between total and viable cell counts.

Cell death occurs by two possible mechanisms : apoptosis and necrosis. Apoptosis is the programmed cell death and is a normal physiological phenomenon. During apoptosis endogenous endonucleases are activated to cleave the DNA into fragments in the form of ladder. The cell membranes have ruffled appearance with many blebs followed by shrinkage of cells, nuclear condensation and fragmentation of the cell into discrete membrane enclosed apoptotic bodies. *In vivo* apoptotic bodies generated by cell breakdown are phagocytosed. In cell culture apoptosis occurs due to depletion of nutrients. In necrosis, cell death occurs due to injury. There is no cell fragmentation characteristic of apoptosis and loss of cell viability is relatively slow. Necrosis occurs in stress condition and characterized by break down of plasma membrane, swelling and rupture of the cell.

Contamination : The main source of contamination is airborne and arises from inadequate aseptic techniques. It mainly originates from hands, breath or hair. The following points should be considered to prevent contamination.

- Wash hands with antiseptic soaps/gel before and after cell culture. Latex gloves may be worn.
- Limit access to the laboratory during experiment.
- Decontamination working surface with 70% ethanol before and after each procedure.
- Put on UV light for about ½ hour in laminar hood before starting any work.
- Face masks and caps should be used. Perspex front of the laminar hood should form a barrier between operator and cell culture.
- Use sterilized glass pipettes, bottles or presterilized plastic pipettes/flasks etc.

- Use media and serum purchased from a reputed company. The serum is a potential source of mycoplasma. Mycoplasma are Gram negative bacteria (0.3-0.5 um) capable of growing inside the cell cytoplasm.

The contamination of media can be checked by inoculation in a variety of broths or nutrient plates. Two commonly used isolation media to reveal the bacterial and fungal contamination are fluid thioglycollate medium and soybean casein digest. Samples are inoculated and kept at 37°C for 7 days.

8
COMPONENTS OF MEDIUM AND ITS PREPARATION

Cell culture medium : To grow cells in *in vitro*, environment should be as close as possible to *in vivo* condition. But a number of factors are important to grow cells *in vitro* namely substrate, temperature and medium. There is a number of ready made medium available commercially and made the cell culture work easier. Any successful medium consists of isotonic buffer solution containing inorganic salts, amino acids, vitamins, a energy source and various supplements. Usually serum or serum replacement is used as supplement.

Type of media and supplements : Earle's basal medium (BME) is one of the original defined medium and commonly used for adherent (Vero, MDBK, MDCK, BHK-21) and primary cells. Other media developed from BME are Dulbecco's modified Eagle's medium (DMEM) having increased amount of vitamin and amino acids. Ham's F-12 contains a wide range of different components which may be required for certain cell lines. A ratio of 1:1 of DMEM and Ham's F-12 has been found effective

48 | Animal Cell Culture and Virology

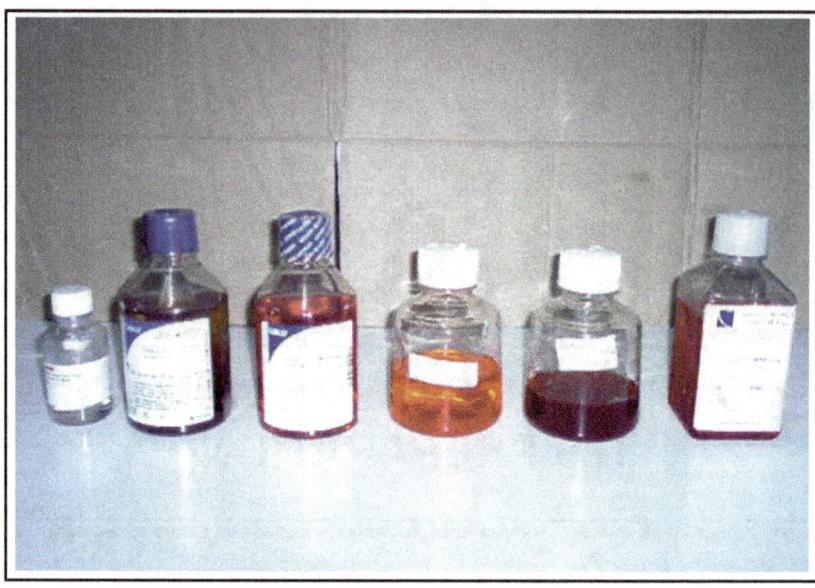

Reading to use media, serum, trypsin-versene solution etc.

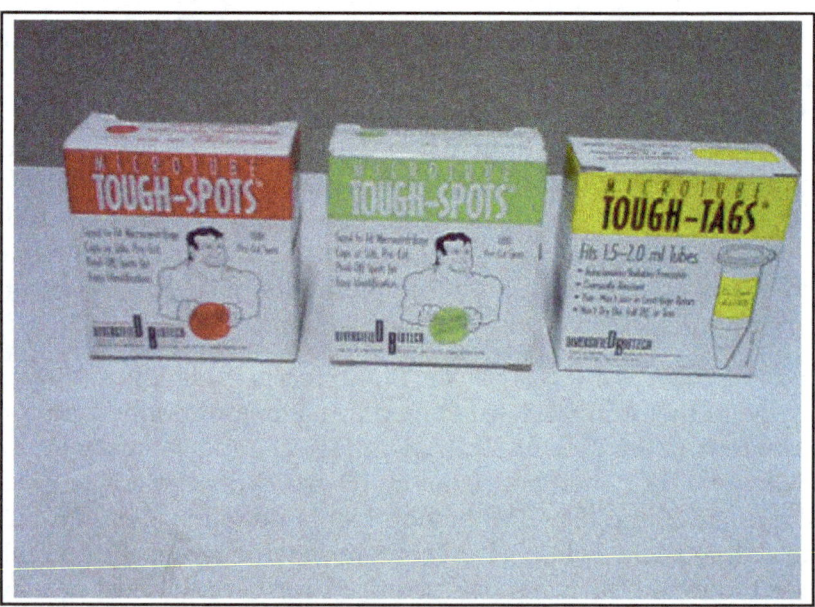

Ready to use label/tag

Components of Medium and Its Preparation 49

as a good basal medium for serum free formulations to support the growth of a number of cell lines. Media can be supplied as 1X liquid ready for use or as a concentrate in liquid or powdered form. The liquid concentrate (10X) is provided sterile and requires dilution with presterilized distilled water. Powdered medium should be dissolved in water and sterilized through a 0.22 μM filter by negative or positive peristaltic pump. The water used to prepare media should be of high purity and be prepared by reverse osmosis or triple distilled water. There are some unstable components (eg. Glutamine, bicarbonate) in culture medium. The half life of glutamine in medium at 4°C is 3 weeks. Liquid medium purchased without glutamine and bicarbonate should be stored at 4°C until used. Additional Ca^{+2} and Mg^{+2} are included in many media preparation and required for adherent protein on anchorage dependent monolayer cell culture. Media for suspension cell culture do not need bivalent cations in high concentration. For example, RPMI (Roswell Park Memorial Institute) 1640 is commonly used to culture non-adherent cells such as human/ murine normal and neoplastic cells and there is a reduced requirement of Ca^{+2}.

The basal medium requires the addition of supplements such as glutamine and serum. Glutamine is the major carbon source for most cells in culture, providing precursors for further biosynthesis and protein production. It is also used as energy source in addition to glucose and Na pyruvate via the Krebs cycle. Serum contains an important but ill defined mixture of growth supporting compounds (polypeptides, hormones, lipids, trace metals etc.). Now-a-days serum like supplement have been available as serum is sometimes unnecessary in some experiments.

Components of a typical culture medium

Carbohydrate : Glucose is an energy source as well as precursor for biosynthesis such as ribose needed in nucleic acid synthesis. Lactic acid is the major product of glycolysis. Only a small amount of glucose is completely oxidized in TCA cycle.

Alternatively fructose can be used and it produces decreased lactic acid and is more stable in culture pH. Glucose can also be replaced by mannose, trehalose, galactose etc. Some cells preferentially utilizes pentoses such as xylose and ribose.

Amino acids : They are used at a concentration of 0.1-0.2 mM except glutamine (2-4 mM) and act as precursor of protein synthesis. The ammonia produced from glutamine either by thermal degradation in the medium or by cellular metabolism is inhibitory to growth in some culture.

Salts : They are included to make the medium isotonic and there is no imbalance with the intracellular content.

Bicarbonate : It is usually included to act as a buffer system in conjunction with 5-10% CO_2 in gaseous phase provided by the incubator. This maintains the pH of culture at 6.9-7.4. In bicarbonate –CO_2 buffer system the culture may become alkaline very quickly when removed from the incubator. To prevent this, an organic buffer called HEPES is used in closed incubator. HEPES (pKa 7.3 at 37°C) is added at the concentration of 10-25 mM. HEPES/bicarbonate combination buffered medium can also be used in a CO_2 gassed system when enhanced buffering capacity is required. HEPES maintains the pH around its pKa value. Other examples of organic buffers are MES (pKa 6.5) and CHES (pKa 9.5). MOPS is another common organic buffer used in cell culture @ 20 mM for nongassed incubators.

Vitamins and hormones : They are present in very low concentration (μmol) and are utilized as metabolic co-factors. The concentration varies considerably from medium to medium as to support the growth of different cell lines.

Phenol red : It is a pH indicator of the medium and become orange at pH 7.0 or yellow at acidic pH or red at slightly alkaline pH or pink at high alkaline pH.

Media supplements

L-glutamine : It is an amino acid which is extremely necessary for cells in culture. It can be purchased as liquid or powder. The powder is dissolved in triple distilled water to make 200 mM, filtered through 0.2 µM filter and kept at -20°C in aliquots until used. The glutamine is added to the medium to a final concentration of 2 mM. It has a relatively short half life, 20% of the glutamine is lost in 3-4 days at 37°C in the medium. If medium stored at 4°C, 20% is lost after 2 weeks.

Serum : Serum is available from most suppliers either with or without heat inactivation. Serum is inactivated at 56°C for 30 min before use to destroy the complement and possible crossreactive immunoglobulins. Serum is used at a concentration of 10% in growth medium to promote cell growth and 2% in maintenance medium to maintain the cells. Bovine or equine serum is most commonly used in cell culture although foetal calf serum (FCS) is the best because of its high content of embryonic growth factor. Other types of serum are new born calf serum. Serum should be sterilized at least once through 0.1 µM filter. Special treatment of serum included :

(1)γ irradiation (2) heat inactivation (3) γ globulin removal (particularly useful in hybridoma culture because it could interfere with the extraction or assay of the antibody) (4) dialysis against 0.15 M NaCl. It reduces the low molecular weight components of the serum but maintain osmolarity. It eliminates carbohydrate, amino acids, nucleotide content in order to study nutrient utilization of cells in culture.

Antibiotics : Penicillin, streptomycin and gentamycin are added @ 100units, 100µg and 50µg per ml of media respectively. Gentamycin is effective against Gram -, Gram + and mycoplasma organisms. Antibiotics are added to the medium at a concentration inhibitory/lethal to the bacteria and mycoplasma but non-toxic to the cells. Similarly antifungal agents are added to inhibit the growth of the yeast, moulds and fungi. The details has been depicted in the table.

Details of some of the most commonly used antibiotics

Antibiotic	Storage temp	Life at 37°C	Active against	Working conc.	Mode of action
Gentamycin	4°C	5 days	Gm +, Gm- and mycoplasma	50 mg/L	Inhibit protein synthesis
Penicillin	-20°C	3 days	Gm +	1 lakh units/ L	Inhibit cell wall synthesis
Streptomycin	-20°C	3 days	Gm + and Gm -	50-100 mg/L	Inhibit protein synthesis
Tylosin	-20°C	3 days	Gm +	6-10 mg/L	Interfere with protein synthesis
Nystatin	-20°C	2 days	Yeast and mould	10-200U/ml	Interfere cell permeability

HEPES : It provides an alternative buffering system that can be used in a non-gassed incubator. It is used at the final concentration of 25mM. In some cases NaCl concentration is reduced to maintain the bicarbonate concentration to be maintained without compromising the osmotic balance of the medium. In non-gassed incubators, bicarbonate will be lost from the culture as CO_2 which affects the growth of cells. As it not only has the buffering capacity but also other functions. HEPES/bicarbonate can be used in CO_2 incubator system when enhanced buffering capacity is required. MOPS, another organic buffer is used in addition to bicarbonate in non-gassed incubator @ 20mM.

Serum free medium : Although serum has been widely used as a medium supplement to promote the growth of cells, there are number of widely recognized disadvantages :

(1) It is chemically undefined and batch to batch variation causes inconsistent cell growth.

(2) It is very much expensive and account for 70-80% of the cost of some formulations.

(3) The proteins in serum can compromise the extraction and purification of cell secreted proteins.

(4) Serum is vulnerable to contaminate with infectious agents *viz.*, bacteria, virus, mycoplasma etc.

In the event of so many disadvantages, efforts have been made to develop supplements of hormones and growth factors to replace serum. Commonly used ingredients for these supplements include insulin, transferrin, ethanolamine and selenite. **Iscoves** medium is a serum free medium prepared by adding BSA, transferrin and soybean lipid to DMEM. Supplement kits can also be obtained from companies like Gibco, Sigma etc. Other serum free media are **AIMV** medium from Gibco and serum and protein free media from Sigma. Serum replacement factors kits like **Ultroser G** or **Ultroser I** or **Ultroser HY** from Gibco for culture of adherent or nonadherent cell are also available. It is always better to reduce the level of serum for culturing cells gradually in order to adapt the cells before switch over to serum free medium.

Serum free formulation used in hydridoma

DMEM/F-12 (1 : 1 V/V)

Insulin : 5 µg/ml

Transferrin – 35 µg/ml

Ethanolamine : 20 µM

Sodium selenite : 1 nM

Courtesy : Murakami, H et al., (1982). Proc, Natl. Acad. Sci. 79 : 1158.

Preparation of Media

Liquid media : Liquid media are available in the form of 10X concentrate or 1X form. The 10X form should be diluted to 1 X with sterilized water. Bicarbonate solution added to the medium varies from medium to medium (given in the table) and finally pH is adjusted to 7.0 -7.2 with either 1 M NaOH or 1 M HCl. Add glutamine and antibiotic solution as per the standard protocol.

Sodium bicarbonate to be added in different media

	g/litre
MEM with Earle's BSS	2.2
Medium 199	2.2
RPMI1640	2.0
DMEM	3.7

Characteristics of commonly used media

BME (Eagle's basal medium)	Eagle's basal medium used for L and Hela cells
EMEM (Eagle's minimum essential medium)	It is used for culturing a number of primary cell and continuous cell lines
DMEM (Dulbecco's modification essential medium)	It is having 4X concentration of amino acids and vitamins of BME
GMEM (Glasgow's modification Eagle's medium)	It is having 2X concentration of amino acids and vitamins of BME
RPMI 1640 (Roswell Park Memorial Institute 1640)	It is used mainly for lymphocyte culture and hybridoma technology.
M-199	Extremely complex medium (61 compounds) and can support cell growth without serum
Leibovitz	In the absence of CO_2 enriched atmosphere, it supports the growth of fibroblasts.

Media preparation from powder : In a large container, 80-90% of the final volume of the medium is filled with sterile triple distilled or deionised water and stirred magnetically. While the water is stirring, slowly add the entire powder, sodium carbonate, sodium pyruvate (needed to maintain some cell line such as PK-15), antibiotic and finally sterilized by filtering through 0.2 µM filter. The filtered media is distributed in different bottles as per requirement and kept at 4°C until used.

Components of Medium and Its Preparation

Autoclavable medium : A few types of media are available from company like Sigma which can be autoclaved. It is made by removing the heat labile elements which are added after autoclaving. The autoclaving is carried out at 121°C, 15 lbs pressure for 15 min. After sterilization, the buffering solution is added aseptically and pH is adjusted to 7.0-7.2.

Preparation of medium : Besides Eagle's medium, Dulbecco's medium, M199, Glasgow's medium, RPMI-1640 are also used for maintenance of cell culture. To make 1000 ml medium, 889 ml of basic medium, 10 ml of 200 mM glutamine, 1ml of gentamycin (50mg/ml) and 100 ml of FCS are mixed.

Materials required
- E-MEM powder for making 1 litre media
- Penicillin – 100 units /ml
- Streptomycin – 100 µg/ml
- Sterile sodium bicarbonate solution (7.5%)
- Glass distilled water
- L-glutamine
- Disposable filter
- Negative pressure unit

Procedure to prepare media
- Disinfect the working hood with an appropriate disinfectant.
- Measure 900 ml of glass distilled water with the help of a measuring cylinder and pour into a 1 litre conical flask.
- Dissolve the E-MEM powdered media in the water.
- Add L-glutamine to the medium (5 ml of 200 mM of L-glutamine)
- Add 1 lakh units of Penicillin and 1 lakh µg of Streptomycin powder per litre of media. Alternatively 1 ml of gentamycin (50 mg/ml) can be used.

- Stir it on a magnetic stirrer until all the ingredients are dissolved.
- Add the 7.5% $NaHCO_3$ drop by drop and adjust the pH at 7.2 and volume upto 1000 ml.
- Filter it with a disposable filter (0.22µM), collect the media in the sterile container and stored at 4°C until further use.
- Add 1 ml each of media to serum agar plate and thioglycolate broth tubes and incubate at 37°C for one week.
- No growth in the serum agar and thioglycollate broth tubes indicates the media is perfectly alright to be used in the cell culture work.

9
PRIMARY CELL CULTURE

Primary cells are often preferred to established cell lines as they better represent the cells *in vivo* in terms of activity and functions. On the other hand, there is significant difference between original culture and present culture and even between cultures of the same cells in different laboratories. The rate of growth increases and there is change in chromosome numbers of lines over the time. The differences in phenotypes can cause problems in comparison and interpretation of results. The advantage of cell lines are ease handling, grow continuously, large yield and readily available. Primary cell cultures compared to established cell line are more fastidious in their growth requirements because many cell lines are derived from tumour tissue and demonstrate reduced growth factor requirements. Some cells (CEF) require little attachment factors or growth promoting factors. Some cell types do not proliferate in culture (rat hepatocytes) where some cells proliferate up to 4 months provided supplied with attachment factors and growth supplements (fetal endothelium). Different supplements required by different primary cells are attachment fators, collagen (type I, II and IV), fibronectin and gelatin for satisfactory growth. Hormones and growth factors include insulin, insulin like growth factor II (IGF-II), interleukin-2 (IL-2), interleukin-3, (IL-3), granulocyte/macrophage colony stimulating factor (GM-CSF), endothelial cell growth supplement (ECGS), parathyroid hormone, fibroblast growth factor (FGF), epidermal growth factor (EGF), hydrocortisone, follicle stimulating hormone (FSH), nerve growth factor (NGF), estrogen and testosterone. Each primary cell requires its own cocktail of factors.

Primary bovine kidney cell culture

Purpose : To do the primary culture taking kidney cells of bovine.

Materials required

- One calf below one year of age
- HBSS
- EMEM
- Haemocytometer
- Coverslips
- Pipettes
- Eppendorf (1.5 and 0.5 ml) tubes
- Microscope
- Trypan blue (0.4%)
- Serum
- Antibiotic solution
- L-glutamine
- 70% ethanol
- Centrifuge tubes
- Trypsin-versene solution
- Scissors and forceps

- Beaker
- Trypsinization flask
- Tissue culture flasks (175 cm² / 75 cm²)
- Magnets
- Muslin cloth
- Centrifuge

Protocol

- Clean the laminar hood with cotton/tissue paper soaked with 70% ethanol and sterilize the hood for about 30 min by putting on the UV light.
- A new born calf below one month of age is sacrificed. Disinfect the surface on the abdominal cavity with 70% ethanol.
- Open the abdominal cavity by making an incision with the help of surgical blade on the abdominal wall to expose the internal organs.
- Remove the kidneys using scissors and forceps aseptically.
- Transfer the kidneys to a sterilized beaker containing HBSS with antibiotics and bring to the laboratory.
- Wash the kidneys with HBSS several times and transfer the kidneys from beaker to a petridish.
- Wash the kidneys several times with HBSS.
- Remove the cortical portion of kidneys and transfer to a petridish containing HBSS. Cut the tissue into smaller pieces of 2-3 mm with scissors to wash the tissue pieces with HBSS and once with trypsin (0.25% of Difco 1:250 trypsin) made in Ca^{+2} and Mg^{+2} free buffer warmed at 37°C.
- Wash again with HBSS to remove RBC, blood clot and other debris.

- Transfer the tissues to the trypinization flask containing sterile teflon coated magnetic bar and sufficient quantity of trypsin (1:10 ; tissue : trypsin).

- Place the flask on a magnetic stirrer for trypsinization and run it at low speed to avoid bubble formation. Discard the supernatant after 10-15 min to remove the toxic substances.

- Pour fresh trypsin and continue trypsinization for 30 min. At the end of each period, allow the tissue to settle down for a few min and collect the supernatant containing cells and passed through a sterile muslin cloth into a beaker and store at 4ºC.

- Again pour the fresh trypsin solution and continue the process 4-5 times.

- Centrifuge the cell suspension at 1000-1500 rpm for 10 min in a refrigerated centrifuge and discard the supernatant. Wash once with growth medium. Resuspend the cell suspension in growth medium containing 10% FCS to get the cell concentration of about 1×10^6-5×10^6 cells/ml.

- Distribute the cell suspension in different flasks.

- Clean the laminar hood with cotton/tissue paper soaked with 70% ethanol and sterilize the hood for about 30 min by putting on the UV light.

- Incubate the vessels at 37ºC in the incubator.

- Observe the culture 24 h interval. A confluent monolayer is formed in 3-4 days or subcultured as per requirement.

Chicken embryo fibroblast culture

Purpose : To do the primary culture taking chicken embryo fibroblast cells.

Materials required
- Ten days old embryonated chicken eggs.
- HBSS
- EMEM
- Haemocytometer
- Coverslips
- Pipettes
- Eppendorf (1.5 and 0.5 ml) tubes
- Microscope
- Trypan blue (0.4%)
- Serum
- Antibiotic solution
- L-glutamine
- 70% ethanol
- Centrifuge tubes
- Trypsin-versene solution
- Scissors and forceps
- Beaker
- Trypsinization flask
- Tissue culture flasks (175 cm^2 / 75 cm^2)
- Magnets
- Muslin cloth
- Centrifuge

Protocol
- Clean the laminar hood with cotton/tissue paper soaked with 70% ethanol and sterilize the hood for about 30 min by putting on the UV light.
- Collect 10 day old embryonated fertile eggs from a disease free flock, candle the egg and select an egg with active movement of embryo with good blood supply.
- Sterilize the surface of the egg and put the egg on an egg tray keeping the air sac at the top most position.

- With the help of sterile forceps and scissors break the egg shell at the air sac, flame the scissors and forceps in between the operation.
- Remove the shell, shell membrane and CAM.
- Take out the embryo with the help of a curved forceps and put it in a beaker containing HBSS with antibiotics.
- Wash the embryo several times with HBSS containing antibiotics.
- With the help of scissors and forceps remove head, limbs and internal organs with scissors.
- Transfer the embryo to a fresh petridish and wash it several times with HBSS to remove RBCs, blood clots and debris.
- With the help of a curved scissors and forceps mince the embryo.
- Wash the minced tissue several times with HBSS, allow the tissue fragments to settle and discard the supernatant and once with trypsin prewarmed at 37°C.
- Transfer the tissue pieces to a trypsinization flask containing sufficient quantity of trypsin and teflon coated magnetic flea.
- Place the trypsinization flask on a magnetic stirrer and run it at low speed.
- After 15 min of trypsinization, allow the tissue fragments to settle, discard the supernatant to remove the toxic substances released from tissue fragments.
- Pour fresh trypsin solution, continue trypsinization for 30 min, allow the tissue to settle and pour the tissue suspension to a beaker through muslin cloth similarly.
- Pour the cell suspension in a centrifuge tube and centrifuge at 1000-1500 rpm for 10 min. Wash the cell suspension with HBSS two times and once with growth medium.

Primary Cell Culture 63

- Suspend the cell in growth medium containing 1% FCS to get the cell concentration of 1×10^5 cells/ml.
- Distribute the cells in culture vessels.
- Clean the laminar hood with cotton/tissue paper soaked with 70% ethanol and sterilize the hood for about 30 min by putting on the UV light.
- Incubate at 37°C in the incubator.
- Confluent mononlayer will be formed in a day.

10

MAINTENANCE OF CELL CULTURE

Cell growth and cell passage : Different types of cells have different growth requirements but in all cases the surrounding medium must supply all the essential nutrients. The temperature, pH, osmolarity and humidity must be kept within limits. Again, toxic or inhibitory substances must not be allowed to accumulate. Serum, a source of macromolecular growth factors, is essential to the growth of many cell types. Anchorage dependent cells require nontoxic or biologically inert glass or plastic surface for attachment.

Upon attaching to the new culture there is no increase in seeding, cells enter a lag phase during which there is no increase in cell number followed by a log phase with increased metabolic activity and exponential increase in cell number. When the nutrients are depleted and toxic products are accumulated, it is called stationary phase, when cell numbers remains constant. It happens when confluent monolayer is formed and no space is left. Some cells are sensitive to contact inhibition while others such as tumours cell are not. Cells become adversely affected and will eventually die if waste products are accumulated. Cells should be passaged regularly to maintain the viability and optimum cell culture conditions.

The cells can be mechanically scraped from the surface or treated with a Ca^{++} and Mg^{++} free enzyme solution. Trypsin is most commonly used although collagenase or pronase can be used. As the enzyme is damaging to the cells, the contact should be minimum. EDTA, a chelating agent is sometimes used along with the trypsin.

Cells grow in suspension culture need not be trypsinized. The cell suspension is centrifuged, the spent medium is discarded and the cells are resuspended in a small amount of fresh medium and put into new culture vessels. When passaging cell cultures, one can put given number of cells or split into 1:3 to 1:4. A 1: 3 split ratio means original cell suspension to be divided into 3 same type of flask/bottles.

Subculturing of cells : The growth medium is decanted from the culture bottles. The cell sheet is washed with trypsin versene solution @ 1 ml per 25 cm^2 bottle for two times. When the cells start to detach from the glass or plastic surface, 2 ml growth medium is added to the bottle. The cell suspension is mixed gently and carefully with the help of an 1 ml pipette to get the single cell suspension. The cell suspension is distributed to 2-3 plastic bottles as per the requirement of the cell bottle in the subsequent days. The cell bottles are incubated at 37^0C in a closed incubator or CO_2 incubator. The confluent cell monolayer will be formed within 2 to 3 days.

Purpose : To maintain the cells in the laboratory by propagating into new culture bottles/flasks.

Materials/preparation
- 25 cm^2 flask of cells (BHK21)
- 25 cm^2 new flasks.

Maintenance of Cell Culture

- EMEM with 10% FCS
- Pipettes
- 37°C water bath
- Trypsin-versene solution (0.25%)

Protocol

- Clean the laminar hood with cotton/tissue paper soaked with 70% ethanol and sterilize the hood for about 30 min by putting on the UV light.
- Observe the confluent monolayer of BHK21 cells under the inverted microscope.
- Remove EMEM with 10% FCS from the refrigerator and warm it at 37°C in water bath/incubator.
- Aseptically decant the spent medium from the flask into the discard bowel inside the hood.
- Add 1 ml of trypsin-versene solution to the cell monolayer and rinse the cell monolayer with the trypsin properly and ultimately decant the trypsin. Repeat the process once again.
- Place the flask in the incubator for 2 minutes.
- Gently shake the flask and check under the microscope for cell detachment.
- Add 5 ml of EMEM with 10% FCS and detach the cells by uniform pipetting.
- Add sufficient volume of the growth medium and distribute into the 3 plastic bottles.
- Cap the flasks and tilt gently to distribute the cells uniformly.
- Clean the laminar hood with cotton/tissue paper soaked with 70% ethanol and sterilize the hood for about 30 min by putting on the UV light.

- Label the flasks and keep inside the incubator at 37°C for cell multiplication and monolayer formation.

Culturing continuous cell lines : There are two type of continuous cell lines i.e. adherent (monolayer cells) and non-adherent (suspension cells). The adherent cells attach to plastic surface i.e. flask or plate and need to be detached for culturing. They derive from organs like muscle, kidney, liver, nerve cells etc and are not motile. On the other hand, non-adherent cells do not generally attach to plastic surface and there is no need to be detached for culturing. All the suspension cultures are derived from immune cells or precursors of immune cells. They circulate within the blood stream and do not generally attach (B and T cells). The adherent/suspension cells can be primary, immortal or transformed.

Contamination and curing : In the cell culture, we usually get the contamination of bacteria, fungi, mycoplasma and yeast. The bacterial contamination can be checked by using antibiotics i.e. penicillin, streptomycin, kanamycin and gentamycin. The mycoplasmal contamination can be inhibited by using gentamycin. The fungal and yeast contamination is very difficult to check and spread via spore motility through air. It is advisable to discard the bottles contaminated with fungi and yeast and take the fresh cells. However, fumigation of the incubator/hood, change of water of CO_2 incubator and use of antimicrobial agent and swabbing the surfaces of the incubator with 70% alcohol can greatly reduce the problem.

Fumigation : To check the serious contamination problem, fumigation is undertaken. Potassium permanganate (5g) is taken in a glass beaker inside the hood or incubator and 50 ml of HCHO is added. The incubator or hood should be closed immediately and sealed properly with tape. After an overnight incubation, the beaker can be removed and its contents disposed of properly. A baker containing 5 g of ammonium carbonate is placed inside the equipment for 2-4 hours to remove last traces of HCHO.

Incubation : For optimal growth most animal cells require a temperature of 37^0C in medium at pH 7.0-7.2. The by-products of cell metabolism tends to make medium pH more acidic than desired. So, a buffering system is required. Most commonly used buffering system bicarbonate/CO_2 system requiring continuous supply of CO_2 (CO_2 incubator). Some laboratory use morpholinopropane sulfonic acid (MOPS) or N-2-hydroxy ethylpiperazine N'-2 ethane sulfonic acid (HEPES) in ordinary dry non-gassed incubator. CO_2 incubator maintains 5%CO_2/95% air at 37^0C with a water reservoir giving high humidity.

11

PACKAGING AND TRANSPORTATION OF CELLS

In may cases, it becomes necessary to send the cells from one laboratory to another laboratory within or outside the country. The ideal method is to fill the flask containing cells at log/exponential phase with growth medium, tightly close the cap, seal it with parafilm and tape and pack it properly with polythene containing tissue paper and cotton. The tissue paper and cotton is used to soak the medium in case of accidental breakage. The polythene containing tissue culture flask having the adherent cell sheet should again be packed with thick paper. Finally, it is to be kept in a wooden/metal box or container. This pack must accompany the letter which provides the detailed information of the cell viz., name of the cell/cell line, passage number, culture medium to be used, split ratio, percentage of FCS to be used, any other supplements to be used and any other condition to maintain the cell in the laboratory. The packed cell is now ready for dispatch at room temperature through post or messenger. Upon receiving the flask, collect the growth medium, shake the culture gently in 2-3 ml medium to remove the loosely adherent cells and discard the medium with dead cells. Add 5 ml of growth medium and keep at 37°C in the incubator. After 1-2 days, when the monolayer becomes confluent, subculture the cells and distribute into the new flask as per requirement.

12
DESCRIPTION OF COMMONLY USED CELL LINES

Vero (monkey, African green, kidney)

Morphology: Fibroblast

Species:, African green monkey

Tissue: Kidney

Properties: Virology, virus titration, virus replication, plaque assays, bacterial cytotoxicity

Susceptible to: Pox virus, bluetongue virus, rinderpest virus, PPR virus, poliovirus 3, echovirus, reovirus, rubella, herpes simplex, arbovirus, adenovirus 12, influenza, paramyxovirus, orthomyxovirus, African swine fever virus, bovine leucosis virus, orbivirus, porcine epidemic diarrhea virus

Karyology : Aneuploid.

BHK-21 clone 13 (hamster, golden Syrian, kidney)
Morphology: Fibroblast

Species: Hamster, golden Syrian

Tissue: Kidney

Properties: Virology, virus replication, virus titration, plaque assays.

Susceptible to: FMD virus, Pox virus, bluetongue virus, Pseudorabies, rubella, vaccinia, herpes simplex, vesicular stomatitis (Indiana), reovirus 3, adenovirus 25.

Karyology : Pseudodiploid.

MDBK (Madin Derby bovine, kidney)
Morphology: Epithelial-like

Species: bovine

Tissue: kidney

Properties: Virology, virus replication, virus titration, plaque assays.

Susceptible to: vesicular stomatitis, parainfluenza 3, infectious bovine rhinotracheitis, bovine diarrhea virus, alpha viruses.

Karyology : Aneuploid.

MDCK (dog, cocker spaniel, Madin Derby canine kidney)
Morphology: Epithelial-like

Species: Cocker spaniel female dog

Tissue: Kidney;

Properties: Virology, virus replication, virus titration, plaque assays.

Susceptible to: Canine parvovirus, adenovirus 4, adenovirus 5, avian influenza virus, influenza A, influenza B, vaccinia, vesicular stomatitis (Indiana), swine fever virus, reovirus 3, reovirus 2, Coxsackie B5, canine hepatitis, canine distemper virus, swine vesicular exanthema virus

Karyology : Aneuploid.

PK 15 (pig, kidney)

Morphology: Epithelial-like

Species: Pig

Tissue: Kidney

Derived from: PK-2a.

Viruses: contains type C retrovirus

Properties: Virology; virus replication, virus titration.

Susceptible to : Swine fever virus, adenovirus 4, adenovirus 5, pseudorabies, vaccinia, vesicular stomatitis (Indiana), reovirus 2, Coxsackie viruses.

Karyology: Aneuploid

RK 13 (rabbit, kidney)

Morphology: Epithelial-like

Species: Rabbit 5 weeks old

Tissue: Kidney.

Viruses: contains bovine diarrhea virus

Properties: Virology, virus replication, virus titration.

Susceptible to: Rabbit pox, pseudorabies, rubella, vaccinia, herpes simplex, infectious bovine rhinotracheitis, simian adenovirus, , myxoma, equine viral arteritis virus

Karyology: Unknown ; non tumorigenic

HeLa (human, black, cervix, carcinoma, epitheloid)
Morphology: Epithelial-like

Species: Black female 31 years old lady

Tissue: Cervix

Tumor: Carcinoma, epitheloid

Properties : Antitumour testing, transformation, tumorigenicity, cytotoxicity, cell biology, bacterial invasiveness, virology

Susceptible to: Adenovirus 3, measles, poliovirus 1, echovirus, vaccinia, arbovirus, respiratory syncytial virus, reovirus 3, rhinovirus, Coxsackie

Karyology: Aneuploid ; tumorigenic in nude mice

Hep 2 (human, Caucasian, larynx, carcinoma, epidermoid)
Morphology: Epithelial-like

Species: Caucasian male 56 years old man

Tissue: Larynx

Tumor: carcinoma, epidermoid

Properties: Virology, virus replication, virus titration.

Susceptible to: Adenovirus 3, poliovirus 1, herpes simplex, vesicular stomatitis (Indiana), respiratory syncytial virus

Karyology: Aneuploid ; tumorigenic

B95a (monkey, marmoset, B lymphoblastoid cell line)
Morphology: Epithelial

Species: monkey, marmoset

Transformed by: Epstein-Barr virus

Derived from: B95-8

Viruses: contains Epstein-Barr virus

Properties: Virology , virus replication, virus titration

Susceptible to: PPR virus, canine distemper virus, measles virus

WI-38 (human, Caucasian, lung, embryonic)
Morphology: Fibroblast-like

Species: Caucasian female lady

Tissue: Lung, embryonic

Viruses: Contains rhinovirus A

Properties: collagen production, biochemistry, vaccine development, protein, virology

Susceptible to: Rabies virus, herpes simplex virus 1, poliovirus 1, vesicular stomatitis, pseudorabies, rhinovirus, rhinovirus (Indiana)

Karyology: Diploid except at high passage number

MRC-5 (human, lung, embryonic)
Morphology: Fibroblast-like

Species: Male 14 weeks old boy

Tissue: Lung, embryonic

Properties: Vaccine development, senescence, virology

Susceptible to: Rabies, measles, poliovirus 1, poliovirus 2, poliovirus 3, echovirus, rubella, herpes simplex, vesicular stomatitis (Indiana), adenovirus

Karyology: Diploid.

CHO (hamster, Chinese, ovary)
Morphology: Fibroblast-like and Epithelial-like

Species: Female Chinese hamster

Tissue: Ovary;

Properties: Virology, genetics, toxicity screening, nutrition, gene expression

13

SUBCULTURING OF CELLS IN DIFFERENT VESSELS

If the concentration of cells in suspension cell culture has reached to 7×10^5 cells/ml, it is to be subcultured. Suspension cells are diluted to concentration of approximately 1×10^5 cells/ml although it varies from cell line to cell line. So, to subculture into a 3 litre glass flask using 500 ml of suspension of cells, there is a need of 5×10^7 cells. As the concentration of cells is 7×10^5 cells/ml, there is a need of 71.4 ml of cells. Adherent cells are put in the culture flasks at a concentration of 5×10^4 cells in a 25 cm^2 flask. For example, after trypsinization, the cells have been resuspended in 5 ml of medium and cell concentration is 1×10^5 cells/ml or total number is 5×10^5 cells. If one wants to put cells into two 25 cm^2 flasks each containing 5×10^4 cells in 5 ml volume, there is a need of 1×10^5 cells, which can be obtained from one ml of cell suspension (1×10^5 cells/ml). While keeping the cells in the CO_2 incubator, the caps of the tissue culture flasks should be loosened to allow the gas to enter the flask. Conversely, when removing a flask from the incubator, caps should be tightened. On the other hand, in non-gassed incubator, tissue culture flasks should be firmly tightened to prevent the evaporation of medium to become alkaline.

Concentration of cells in different types of culture vessels

Type of vessel	Growth area cm^2	Adherent cell number X10^5	Medium volume	Suspension cells X 10^5
24 well plate	2	0.04	1ml	1-2
6 well plate	9.6	0.2	3-4 ml	3-6
25 cm^2 flask	25	0.5	5-10 ml	10-20
75 cm^2 flask	75	1.5	15-30 ml	30-60
175 cm^2 flask	175	3	50-60	-

Adult or embryo : The primary cell cultures are often obtained from embryo tissues as they survive and proliferate better than tissues from adult. Embryonic or young animal tissues are having lower level of specialization and higher proliferative potential whereas reverse is true for adult tissues.

Normal or neoplastic : Normal tissue usually gives rise to cultures with finite lifespan, while cultures from tumours can give rise to continuous/transformed cell lines. However, there are examples of continuous cell lines (BHK-21, MDCK, MDBK, Vero etc) that have derived from normal tissue and are non-tumerogenic. The culture of cells taken directly from animal is called primary culture until they are subcultured. They are usually heterogenous and have a lower growth rate.

Finite or continuous cell lines : After several subcultures, a cell line will either die out (finite cell line) or transform to become a continuous cell line. Compared to finite cell culture, indefinite or continuous cell lines show immortalization due to mutation and/or deletions of anti-tumour gene p53, higher telomerase activity and indefinite life span. The continuous cell line is characterized by an alteration of cytomorphology (smaller, less adherent, more rounded and higher nucleus to cytoplasm ratio), increase in growth rate, reduction in serum dependence, an increase in cloning efficiency, a reduction in anchorage dependence (ability to proliferate in suspension), an increase in heteroploidy and aneuploidy and increase in tumerogenicity. The resemblance between spontaneous *in vitro* transformation and malignant transformation is obvious but two are not identical. Normal cells can transform to become continuous lines without being malignant whereas malignant tumours can give rise to culture which transform and become tumerogenic. The advantages of continous cell lines are faster growth rate to higher cell densities, lower serum requirement and ability to grow in suspension. The disadvantages are greater chromosomal instability, divergence from the donor phenotype and loss of tissue specific markers. A number of techniques include transfection or injection with viral genes (E6 and E7 from human papilloma virus and SV40 T) or viruses (EBV) have been used to immortalize a wide range of cell types.

Suspension cultures : Most of the cultures are propagated as a monolayer attached to the substrate but few such as transformed cells, haematopoitic cells and cells from ascites can be propagated in suspension. The advantages of suspension culture include simple propagation (no trypsinization), no need of increase surface area, ease of harvesting, inexpensive and possibility to achieve a steady state.

14
SCALING UP OF ANIMAL CELL CULTURE

Small volume cultures are useful for experiments such as studying cell morphology, titration of viruses, plaque assay, pock assay, SNT and to compare the effect of agents on growth and metabolism. However, there are other application where there is a need of large number of cells such as extraction of DNA/RNA from cells, production of viruses for vaccine production and a number of products namely interferons, interleukins, hormones, enzymes, antibodies, glycoproteins etc. The large quantity of cells can be obtained by using a large number of flasks/bottles but it is tedius, labour intensive, expensive and there is a risk of contiamination. Although the unit processes are more cost effective and efficient, it needs some modification to overcome limiting factors such as oxygen depletion, shear damage and metabolic toxicity. There are a number of aberrations available to scale up cell culture described below.

(1) Roller bottles

When large volumes of cells or culture fluid is required, adherent cells are cultured in roller bottle. A small roller rack can be placed in the incubator or a larger one inside the walk in incubator. Sterile disposable roller bottles are of two types (solid and vented) and available in different sizes. Bottles with vented cap must be kept inside the CO_2 incubator. The cell density should be relatively high (2 to 4 fold more cells /cm^2 surface area) than in flask/plate as relatively reduced initial growth rate of many cells in roller bottles due to decreased control over pH if not grown in a CO_2 incubator. It is better to use an organic buffer (HEPES @ 20 mM) and an air space containing 5% CO_2. The level of the roller bottles should be perfect from front to back to prevent any uneven plating of cells. The roller bottles are rotated initially at a slow speed (0.25 rpm) until cells are attached. Thereafter, a faster speed is set although most of the cells are rotated at 0.5-1 rpm. If the cells are off the bottle in sheets, reduce the speed. After every complete revolution of the bottle, the entire cell monolayer has transiently been exposed to the medium. The volume of medium need only be sufficient to provide a shallow covering over the medium. The cell culture infected with virus or for production of any product should be harvested when the titre of virus or the concentration of the product is high. If the cells are tightly attached to the surface, special scrapers designed to get cells out of roller bottle can be used.

Spinner cultures : Spinner cultures are used for scaling up the production of non-adherent cells/suspension cells. It is carried out in a glass flask with a central teflon paddle that turns and agitates the medium when placed on a magnetic stirrer. Alternatively Schott Duran bottle containing a sterile magnetic stirring flea can be used. The stirring does not allow the cells to be settled down at the bottom and improves gas exchange. The medium is buffered with HEPES and a speed of 30 rpm is optimal for most of the cell types.

Microcarrier beads : Usually tissue culture flasks are used to grow the adherent cells. However, these are limited in terms of the available area for growth. Microcarriers are good alternative to roller bottles to grow adherent cells having high growth surface to volume ratio. Microcarriers are microscopic particles of about 200 µM that are easily maintained in suspension in liquid medium. The beads can be made from a variety of materials *viz.*, dextran, metal, agarose, collagen, DEAE sephadex, cellulose, gelatin etc. The type of microcarriers selected depends on the cell type, medium and container. The suspension of microcarriers is maintained in suspension in liquid medium. Microcarrier cultures are the most useful for harvesting secreted protein from the medium as the cells and medium are easily separated and beads can be used for further culture. Microcarrier culture is less convenient when cell harvest is desired as the separation of the beads and cells is a laborious process and many cells will be lost in the process. If the cell harvest is desired, silica beads are the best option as cells can be easily separated. If subcellular fraction is desired, direct lysis of the cells on the beads are recommended. The most commonly used microcarriers are made up of dextrans. They are relatively inexpensive, readily available, easily sterilized and most cells stick to them. One disadvantage is that many secreted protein will also bind to the bead materials. Microcarriers are suspended at the concentration to provide 0.24 m^2 for every 100 ml of culture compared with 200 cm^2 in stationary culture flask. Cells growing at such high densities will rapidly exhaust the medium and may need replacement of medium during culture and addition of an organic buffer (HEPES) to assist pH stability. The suspension of microcarriers is maintained in a spinner flask which contains a magnetic flea and a paddle suspended from a stirring shaft. The flask can be 100-500 ml in volume and placed on a magnetic stirrer/microcarrier stirring platform to allow a slow and controlled speed of around 40 rpm. The major advantages are increased available surface area for cell growth, light and allow easy suspension in culture medium, transparent to allow easy visualization of cell attachment and growth and charged to allow easy cell attachment on the surface. Each microcarrier can accommodate 100-200 cells. The cell concentration should be such that there is even distribution of cells on the available microcarriers. The stirring rate should be such that it allows microcarriers to be suspended in the medium but do not damage the cells. Harvesting of cells from microcarriers can be done by treating with proteolytic enzymes (trypsin/collagenase). The advantage of microcarrier culture can be made from the comparison : A Roux flask has a surface area of 0.02 m^2 and can accommodate 10^8 cells whereas 1 g (dry weight) of solid microcarriers in 1 litre of culture provides a total surface area of 0.64 m^2 and can supply 2X10^9 cells (equivalent to 20 Roux flasks). A typical microcarrier culture is established with 1-3 g (dry weight) of microcarrier /litre. This allows a final cell density of around 2X10^6 cells/ml from an initial inoculum density of 2X10^5 cells/ml.

Hollow fibers : This consists of bundles of synthetic seripermeable hollow fibers which offer a matrix for cell growth similar to vascular system *in vivo*. Liquid can flow through fibers (the intracapillary space) or through the space between the fibers (extracapillary space). The culture medium is pumped through the intracapillary space and a hydrostatic pressure permits the exchange of nutrients and waste products across the capillary wall. The cells and large molecules are held in the extracapillary space. In this process, a concentration 10^9 cells ml can be obtained. Medium is pumped into the fibers creating a relatively high pressure within the intracapillary space forcing nutrients through the pores into the extracapillary space. They are very effective for suspension cells at scale up to 1 litre (1.2X10^8 cells). It is appropriate only where culture fluid is desired as it is difficult to recover cells from these systems. The harvest fluid can be collected on regular basis or even online so that large quantities of fluid are not accumulated as with the roller bottle or spinner harvest. It is effective for production of protein which are unstable at culture medium at 37°C as the medium can be harvested on a regular basis. Hollow fiber system are available in different sizes.

Porous carriers : Microcarriers and glass spheres are restricted to attached cells and has a low surface area/volume ratio compared to porous carriers. The advantages of porous carriers over solid carriers are 20 to 50 fold higher cell density and support both attached and suspension cells, immobilization in 3D configuration, suitable for stirred fluidizer and fixed reactors, cells protected from shear and capable of long term continuous culture. They can be used both for anchorage dependent and independent cells. The original dextran microcarriers (Cytodex) are microporous but pore size is not sufficient to allow cells to enter inside the beads. However, macroporous carriers such Cultispher are spherical gelatin microcarriers having pore large enough for cells to enter and grow inside the beads. They have the advantage of increase surface area for cell attachment and cells within the beads are protected from shear force.

Fermentors : They are available in different sizes viz., 1, 2, 5, 10, 20, 50, 100, 200, 500 litre capacities. It is the best way to produce large quantities of cells, viruses for vaccine production or recombinant proteins. There is automated control of temperature, CO_2 level, pH etc. They are mainly used by manufacturing companies for production of various vaccines against different diseases of humans and animals.

Multitray unit/flask : It is a kind of stationary culture used to culture adherent cells. The plastic flasks are available in the form of 3 to 10 chambers to increase the surface area. They are made up of tissue culture grade plastics and disposable. The disadvantage is that after trypsinization all cells are not harvested and unlike tissue culture flasks all the layers of cells can not be visualized under the microscope.

15

CELL LINE AUTHENTICATION AND CHARACTERIZATION

There are a number of different cell lines derived from human and animals. With the dramatic increase in numbers of cell lines, the risk of intraspecies and interspecies cross contamination rises proportionally. It is particularly a major problem in laboratories where many different cell lines of human and animal origin are handled at a time. The purity of the primary cell or continuous cell lines and species of its origin should be checked on a regular basis to authenticate the cell line identity. In the absence of such monitoring, inter and intra species cell line contamination are likely to occur in the laboratory resulting in loss of investigator's time effort and resources. The advantages of working with a well defined cell line free from contaminating organisms would appear obvious. There are numerous instances where cell lines are exchanged between the laboratories and it is a probable cause that cell of one species is contaminated with other species or one cell line is contaminated with other cell line. There are a number of procedures to check the authenticity of cell and described below.

(1) Cytogenetic analysis : The chromosomal content of a cell line examined microscopically provides a method for confirming the species of origin and detection of aberrations in chromosome number and/or morphology. It is very much useful for species identification of cell lines with unique chromosome markers and to differentiate cell lines apparently identical by isoenzyme analysis. However, it is time consuming and full characterization and regular monitoring of identity is impracticable.

(2) *Isoenzyme analysis* : Isoenzymes are enzymes that exhibit interspecies and intraspecies polymorphisms that can be observed by differences in electrophoretic mobility. By using enzyme specific colorimetric staining reactions of electrophoretically separated cell extracts, it is possible to readily determine the electrophoretic mobility of many different isoenzymes in a cell extract. It is useful for species identification of one cell line from another. It utilizes the property of isoenzymes having similar substrate specificity but different molecular structures which affects their electrophoretic mobility. Thus each species will have a characteristic mobility pattern of isoenzymes. While the species of origin of a cell line can usually be determined with only two isoenzyme tests (Lactate dehydrogenase and glucose-6-phosphate dehydrogenase) specific identification of a cell line would require a battery of tests. However, the use of a wide range of isoenzyme tests for accurate identification of cell lines requires a detailed knowledge of each isoenzyme system.

(3) *Giemsa banding (G-banding)* : It is a powerful method for identification of cell by karyotype analysis after treatment with trypsin and Giemsa stain (G-banding). The banding pattern are characteristic for each chromosome pair and permits recognition even of minor inversions, deletions

or translocation by an experienced cytogeneticist. Many lines retain multiple marker chromosomes readily recognized by G banding and serve to identify the cells specifically and positively.

(4) *DNA fingerprinting* : This technique is more commonly used in forensic science but can be used in cell line identity. A unique DNA fingerprint can be developed for a particular cell line. A distinct DNA fingerprint results from fragmentation pattern can be obtained when DNA is digested with one or more restriction enzymes. The resulting restriction fragment digest is separated by electrophoresis. The DNA bands are transferred to the nitrocellulose paper followed by hybridization with a radio-labelled probe and autoradiography. The most useful probes for this purpose are those that hybridize to minisatellite DNA. These are repetitive nucleotide sequence of varying length found throughout the genomic DNA. Certain restriction enzymes particularly Hinf 1 is used because they cut DNA within the minisatellite regions. The length and distribution of minisatellite DNA fragments are unique to individuals and hence can be used for identification of cell line.

16
CRYOPRESERVATION AND THAWING/REVIVAL OF CELLS

Early passage cell lines are relatively unstable and go through a period of adaptation to culture. However, between 5th and 35th generation (human diploid fibroblasts), the culture is fairly consistent although it varies from one cell to another cell. Finite cell lines should be preserved after adaptation but before senescence (getting older). Cells should be preserved to protect cell line instability, availability in case of contamination, incubator failure or accidental loss.

Cryopreservation and thawing : In the cell culture laboratory, a number of cells/ cell lines are maintained but not all are necessary to be grown at the same time. So, it is necessary to cryopreserve the cells. A great amount of time, effort and money would be wasted if cells are not preserved when they are not needed. Cryopreservation involves the storing of cells at -196°C (liquid nitrogen) and would be a ready supply in case of contamination or loss of cells due to other reasons. While storing the cells it should be kept in mind that liquid nitrogen will cause burning due to extreme low temperature and one should wear eye protection and heavy gloves as a measure of caution.

Cryopreservatives : Glycerol or DMSO are most commonly used as cryoprotective agents. The function of cryoprotective is to reduce the water content of the cells. The cryopreservatives are small molecules, soluble in lipids and enter cells by diffusion across the lipid bilayer of cell membranes. There will be no formation of ice crystals which rupture cell membranes in the presence of cryopreservatives. The high serum concentration maintain the cell integrity by maintaining the protein concentration rendered permeable by cryopreservatives. An ideal freezing mixture contains 70% (V/V) medium, 20% (V/V) FCS and 10% (V/V) DMSO or glycerol. The mixture is mixed properly and filtered through 0.22 µM filter and kept at 4°C until used. In this mixture, equal quantity of cell suspension is added for preservation. Cells should be cryopreserved only when they are healthy and in exponential phase. Confluent or overgrown cells should not be preserved as they can not be revived well from cryopreservation. The adherent cells in the flasks are trypnized and make suspension with ice cold medium containing 10% FCS. Adherent cells should be resuspended to 2×10^6 cells/ml or 1×10^7 cells/ml. The cell suspension should be placed on ice and equal quantity of freezing mixture added and mixed thoroughly. Aliquots (1 ml) of cell suspension should be put into chilled cryovials and the cell concentration should be 1×10^6 or 5×10^6 cells/ml. The screw caps of the cryovials must not be overtightened as this will distort the gaskets and cause liquid nitrogen to enter the vials. Usually the cryovials are kept in a polystyrene box sealed properly and placed in a -70°C freezer where the cells will cool down at 1°C/min which is desirable. It will take 3 hours to be ready for transfer to liquid nitrogen container. The cryovials containing cells are collected from liquid nitrogen and put into a beaker containing 37°C water during revival. Care must be taken as liquid nitrogen may burn the skin or the vials can be exploded. The vials from the 37°C water are taken away, wiped with 70% ethanol soaked cotton and taken into the hood. The cells preserved with glycerol can be put into 25 cm² culture flask containing 5 ml growth medium. Next day the medium is discarded and fresh growth medium is added at the rate of 5 ml per 25 cm² flask. However, cell preserved with DMSO should be centrifuged at 150-200 g for 5 min after taking out from liquid nitrogen, supernatant discarded and cell suspension are put into 25 cm² flask containing 5 ml growth medium. Cells recover from liquid nitrogen with differing efficiency should be observed carefully for a number of days and subcultured when necessary. Some lines give a large percentage of dead cells after 24 hours and take 3-4 days to become healthy and confluent while other recover almost immediately.

Cryopreservation : When the cells are maintained at reduced temperature, their metabolic rate is slowed down and require less frequent pasaging. At extremely cold temperature, the destruction processes and normal processes are inhibited. Cells can be stored at -70°C in the presence of cryopreservative but only for few months to year and there is a gradual loss of viability over the time. The cells in the presence of cryoprotective agents in liquid nitrogen can be stored indefinitely. In the absence of cryoprotective agent such as glycerol or DMSO, the damage of cells occurs by mechanical injury due to ice crystal formation, changes in the concentration of electrolytes and pH, dehydration and protein denaturation during freezing process. In order to minimize the damage, glycerol or DMSO is added to lower the freezing point and to protect the membrane from rupture. Further a slow cooling rate allows water to move out the cells before freezing and storage below -130°C retards ice crystal formation when cells are thawed rapidly after taking out from liquid nitrogen to minimize the cell damage during -50°C to 0°C temperature range. Different cell types vary greatly in response to the physical and chemical trauma of freezing and thawing. No single medium or procedure is suitable for all type cells. Usually most cell culture respond well to medium containing 5-10% glycerol or DMSO, cooling carried out @ 1-10°C per min and 5-25% FCS protect cells during freezing.

Purpose : To preserve the cells at low temperature or liquid nitrogen and to revive the cells in case of requirement.

Materials required
- One 25 cm^2 flask of Vero cells
- EMEM
- Cryoprotective agent (glycerol/DMSO)
- Trypan blue for viable count (0.4%)
- Haemocytometer
- Sterile pipettes
- 2 ml cryovials
- Trypsin versene solution
- Discard container
- Sterile 1.5 ml /0.5 ml eppendorf tubes

Protocol
- Clean the laminar hood with cotton/tissue paper soaked with 70% ethanol and sterilize the hood for about 30 min by putting on the UV light.

- Observe the cells in a 25 cm^2 flask subcultured 24 hours before under the microscope. The cells should be healthy not overconfluent but in log/exponential phase.

- Decant the medium and wash two time each with 1 ml trypsin-versene solution prewarmed at 37°C.

- Keep the flask at 37°C for 1-2 min until cells start detach from the surface. Add 1ml growth medium containing 10% FCS and pipet the cells several times to make single cell suspension.

- Take 0.1 ml of cell suspension in an eppendorf tube and add 0.1 ml of trypan blue.

Cryopreservation and Thawing / Revival of Cells | 97

- Load the haemacytometer and calculate the number of viable cells/ml.
- Adjust the cell suspension to obtain 2×10^6 viable cells/ml.
- Add one ml of cell suspension to 1 ml mixture containing 70% EMEM, 20% FCS and 10% DMSO/glycerol. Mixed well and transferred to 2 ml chilled sterile cryovial.
- Transfer the cryovials to 4°C for 1 h, -5°C for 1 h (freezer chamber of refrigerator), -20°C for 1 h, -70/-80°C for 1 hour and finally to liquid nitrogen. The cryovials should bear a label with informations of cell types, passage number and date.
- Clean the laminar hood with cotton/tissue paper soaked with 70% ethanol and sterilize the hood for about 30 min by putting on the UV light.

Note : Some cell lines are adversely affected by prolonged contact with DMSO. This can be reduced by adding the DMSO to the cell suspension at 4°C and removing it immediately upon thawing. If it does not work, lower the concentration of DMSO or use glycerol. Although less toxic to cells than DMSO, glycerol can cause change in osmotic pressure during thawing. It should be added to the cell suspension at room temperature during storage. High serum concentration may help cells survive freezing. The temperature of the vapor phase of liquid nitrogen varies between -140°C to -180°C and the liquid phase is -196°C. The storage of cells in vapour phase is greatly reduces the explosion of the leaky vials during removal.

Thawing/Revival of cells

- Remove the vial from the liquid nitrogen and place in the beaker containing water at 37°C. Gloves and face mask should be worn as leaky vials may explode.
- Thaw the contents of the vial by agitation immediately after thawing.

- Remove the vial from the water and wipe up the outer surface with cotton soaked with 70% ethanol and allow to dry at room temperature.

- Clean the laminar hood with cotton/tissue paper soaked with 70% ethanol and sterilize the hood for about 30 min by putting on the UV light.

- Carefully open the vial inside the hood and aseptically transfer the content to a 25 cm^2 flask containing 5 ml growth medium if glycerol is used as cryoprotective agent. Transfer the cell suspension to a sterile centrifuge tube containing growth medium (1 part cell suspension : 9 parts growth medium) and centrifuge at 200g for 5 min.

- Discard the supernatant and resuspend the cell in 5 ml growth medium and transfer to a 25 cm^2 flask and keep inside the incubator at 37°C.

- Clean the laminar hood with cotton/tissue paper soaked with 70% ethanol and sterilize the hood for about 30 min by putting on the UV light.

- Next day medium is discarded, cell sheet is washed with medium and put 5 ml fresh growth medium in case of glycerol preservative. There is no need to change the medium in case DMSO after 24 hour.

- Confluent monolayer of cell can be obtained after 2-3 days.

17
CONTAMINATION OF CELL CULTURE AND CURING

Cell culture contamination continues to be a major problem and a subject of great concern. Contamination may enter the cell culture system as physical, chemical or biological component of the environment and biological contamination represent the greatest threat as living organism metabolize and replicate. Replication increases the titre of the contaminant.

Physical contamination : Physical components of cell culture systems such as radiation, irradiation (UV light and fluorescent light), temperature extremes can cause reduced cell growth and cell death. To minimize the problems, incubators should be located where ambient temperature is relatively constant and no vibration (near centrifuge). Media and media components should not be stored in glass door refrigerators, next to radioisotopes, bench tops or outside the refrigerator or in the hood for prolonged period.

Chemical contamination : It can enter the cell culture system from many sources such as high concentration of amino acids in the medium, carry over of toxic soap or detergent residues in glasswares etc.

Biological contamination : Cell culture can be contaminated with bacteia, fungi, yeast and mycoplasmas. To prevent the contamination, antibiotics, antifungal and antimycoplasmal agents are added to the culture.

Different media used to detect bacteria or fungi

Medium	Temperature	Aerobic/anaerobic	Observation time
Blood agar	37°C	Aerobic and anaerobic	14 days
Thioglycollate broth	37°C	Aerobic and anaerobic	14 days
Brain heart infusion broth	37°C	Aerobic	14 days
Sabouraud broth	37°C	Aerobic	21
Nutrient broth	37°C	Aerobic	21
2% yeast extract	37°C	Aerobic	21

Bacteria may be seen under the microscope as round or rod shaped and they may be motile. In case of bacterial growth, the culture fluid will be cloudy and yellow in colour. Bacteira will use up the nutrients in the medium and excrete waste products. Cultures with bacterial contamination must be discarded. As a general rule, bacterial contamination results from human touch, improperly sterilized glass wares and contaminated air. Fungi can be visualized by naked eye as long hyphal growth with fuzzy appearance. Fungi usually spread in the air. Yeast is a fungal microorganism with round or ovoid bodies that grows like a branched string of pearls. Home baking/brewing is the source of yeast. Mycoplasma is similar to bacteria but smaller in size and without a cell wall. They grow inside the contaminated cells. It is very difficult to detect as it does not usually change the culture medium and not visible under the light microscope. However, in some cases, vacuoles in the cells or excess extracellular matrix can be seen. Cell line, serum, trypsin etc may be a source of mycoplasma. The mycoplasma can come from the mouth/breath of the operator.

Curing : It is better to discard the cell culture contaminated with bacteria. Fungal and yeast contamination present more difficulties because it spread via spore mobility in the air. Following guidelines should be followed to minimize the contamination in cell culture.

1. The incubator should be cleaned with 70% ethanol.
2. Keep the flasks inside the closed incubator instead of CO_2 incubators when there is frequent problem of contamination.
3. The medium should be checked by incubating at 37°C for 24-48 hours.
4. Clean the hood with 70% ethanol and UV light before and after work.
5. Change the water and antimicrobial agents kept at the base inside the incubator at least 10-15 days interval.
6. Put antibacterial, antifungal and antimicoplasmal agents in the medium as per the recommendation to check the bacterial, fungal and mycoplasmal growth.

Bacteria can be inhibited by broad spectrum antibiotics eg gentamycin or penicillin and streptomycin mixture. It is to be remembered that some cell lines are susceptible to effect of certain antibiotics. The use of gentamycin may be beneficial as it is effective against both bacteria and mycoplasma. The antifungal agents can be used in the medium to check the growth of fungi. However, it is not recommended for long period use because of toxicity to cells. Mycoplasma (0.3-0.5 µM) infects the cytoplasm of mammalian cells and affects cell metabolism. For example, *M. arginini* has a high requirement for arginine and cause rapid increase in culture pH. The best way to ensure a mycoplasma free cell line is to test for mycoplasma contaminations at regular intervals (6 months). Several methods such as isolation of mycoplasma, DNA staining using fluorochrome and immunological or biochemical methods are available to detect mycoplasma contamination. However, molecular methods such as hybridization using rDNA gene probes or PCR have been found convenient and sensitive method for routine detection of cell culture contaminations.

18

COUNTING OF CELLS

Enumeration of cells is essential to determine the multiplicity of infection, efficiency of plating and to obtain the specific number of cells for an experiment. The concentration of cells of a suspension can be counted using a haemocytometer. Each chamber is divided into a grid of nine large squares. Each large square is 1mmX1mm (area 1 mm^2) and the depth is o.1 mm. The volume of each square is 0.1 mm^3 or 10^{-4} cm^3. Since 1 cm^3 is equal to 1 ml, the number of cells per ml of suspension can be obtained by multiply the average count per large square by 10^4 and reciprocal of dilution factor. Dilution is made in such a way that one large square contain about 20-60 cells. The haemocytometer chambers should be filled by capillary action and cell clump are counted as one cell. A total count does not distinguish between living and dead cells. The number of viable cells can be determined by staining the cells with trypan blue or erythrosine B. In trypan blue staining cells appear as unstained whereas dead cells appear as blue.

Purpose : To know the total cell count as well as viable cell count of a cell suspension.

Materials required
- One 25 cm^2 flask containing MDBK cells.
- EMEM
- Haemocytometer
- Coverslips
- Pipettes
- Eppendorf (1.5 and 0.5 ml) tubes
- Microscope
- Trypan blue (0.4%)
- Serum

Protocol
- Clean the laminar hood with cotton/tissue paper soaked with 70% ethanol and sterilize the hood for about 30 min by putting on the UV light.

- Observe the cells in a 25 cm^2 flask subcultured 24 hours before under the microscope. The cells should be healthy not overconfluent but in log/exponential phase.

- Decant the medium and wash two times each with 1 ml trypsin-versene solution prewarmed at 37°C.

- Keep the flask at 37°C for 1-2 min until cells start detach from the surface. Add 1ml growth medium containing 10% FCS and pipet the cells several times to make single cell suspension.

- Take 0.1 ml of cell suspension in an eppendorf tube and add 0.1 ml of trypan blue.

- Mix the cell suspension with the dye with micropipette tips.

Counting of Cells

- Clean the haemocytometer and place a coverslip over the squares. Charge the cell suspension with the help of pipet tips under the coverslip.
- Use low power objective on the microscope to count all the cells for total count and unstained cells for viable count.
- Count the cells touching the right and bottom lines but not the top and left line.
- The number of viable cells in the original suspension is determined as follows :
- Suppose total number of unstained cells in 5 squares (one central and 4 corners) are 165, then

$$\text{Number of viable cells} = \frac{\text{Total number of unstained cells in 5 squares} \times 10^4 \times 2 \text{ (dilution factor)}}{5} /ml$$

$$= \frac{165 \times 10^4 \times 2}{5} = 6.6 \times 10^5 \text{ cells}/ml$$

Suppose total number of cells (both unstained and stained) in 5 squares are 195, then

$$\text{Number of total cells} = \frac{\text{Total number of cells in 5 squares} \times 10^4 \times 2 \text{ (dilution factor)}}{5} /ml$$

$$= \frac{195 \times 10^4 \times 2}{5} = 7.8 \times 10^5 \text{ cells}/ml$$

19
ISOLATION AND IDENTIFICATION OF VIRUS IN CELL CULTURE

Cell culture is the most convenient and widely used system for the isolation and propagation of many viruses. Either primary cell culture or established cell lines can be used to isolate the suspected viruses. One can normally get 10-100 virus particles per cell inoculated. Many details of the infection process are controllable to a high degree (including cell density, age, multiplicity of infection (M.O.I.), medium composition, cell type and length and temperature of incubation). Many cell types from different species can be susceptible to a particular virus and vice versa. For replication to occur, the cell type used must be permissive for a particular virus under study.

Virus growth in cell culture can be monitored through appearance of cytopathic effects (CPE) in cell culture. CPE (rounded cell, granularity, vacuolization, syncytia formation etc.) gradually becomes visible over the time during the replication process.. The formation of inclusion bodies is also evident in virus infected cells. Intra-cytoplasmic (I/C) inclusion bodies are found in cells infected with pox viruses, paramyxoviruses, reovirus and rabies viruses whereas intranuclear (I/N) inclusion bodies are produced by herpes viruses, adenoviruses and parvoviruses. Some viruses *viz.*, canine distemper and RP viruses may produce both I/C and I/N inclusion bodies in the same cell. Inclusion bodies can be shown by HE staining, fluorescent antibody staining or electron microscopy. The basophilic I/C inclusion bodies are found in cell infected with pox viruses.

Cell cultures suitable for isolation of viruses from clinical specimens

Serial No	Viruses	Cell /cell lines used
1.	FMD virus	BHK21, PK15, IB-RS-2, bovine kidney, Hela and Hep2
2.	PPR virus	Vero, bovine kidney, bovine testis and B95a
3.	RP virus	Same as PPR virus
4.	IBR virus	MDBK, WI 38
5.	BT virus	BHK21, Vero, Bovine and ovine embryonic kidney and testis
6.	Canine parvo virus	CRFK, MDCK, canine kidney, feline kidney
7.	Rabies	CEF*, MEF**, pig kidney, dog kidney, hamster kidney, BHK21, WI38, MRC5, Human embryonic lung
8.	Sheeppox and goatpox	Kidney and testis of sheep, goat and calf, MDBK, Vero and BHK21
9.	Swine fever	Kidney, testis, bone marrow, spleen, lymph node of swine, PK15

* CEF = Chicken embryo fibroblast
** MEF = Mouse embryo fibroblast

Another conspicuous feature of infection of cell monolayer by paramyxo, herpesviruses, corona and pox viruses is production of syncytia/giant cells or polykaryocytes.

The non-cytopathic virus can be monitored by (a) synthesis of viral nucleic acid and protein (b) haemadsorption – ability of the cells infected with haemagglutinating viruses to adsorp RBC (c) Immunoflurescence – Fluorescent labeled antibody is allowed to react with infected cells and in the positive case, fluorescence will be observed. (d) Interference – Multiplication of one virus is inhibited by another virus.

20

VIRUS ASSAYS

The most important property of a virus is its infectivity or ability to invade a cell and parasitize that cell to replicate itself. To measure infectivity, a titre is defined as the given number of infectious virus units/unit volume and an infectious unit is the smallest amount of virus that produces some recognizable effect in the host system. There are two basic types of infectivity assays (a) Quantal assays and (b) Quantitative assays.

Quantal assays do not measure the number of infectious virus particles present in the inoculum but denote a value for the virus titre where measurement is based on an all or none principle, CPE + or -, dead or alive etc. In the quantitative assays, the number of virus particles present in the suspension can be quantified.

In quantal assays, serial dilutions of virus are inoculated in (a) cell culture (b) eggs and (c) animals. After a certain period observations are made. At each dilution the animal or test unit is either scored as being infected or uninfected. At the higher virus dilution the hosts are unaffected whereas at the lower dilution all the hosts are affected. The virus titer is calculated using the intermediate dilutions where only some of the hosts are infected, and the titer is expressed as the reciprocal of the dilution of virus that infect 50% of the inoculated host/system using Reed and Muench method. At this virus dilution each infected host would contain 1 ID_{50}. The titre is expressed as $TCID_{50}$, LD_{50} or EID_{50} in cell culture, animal and eggs indicator system, respectively.

Materials required
- 24 well tissue culture plates
- 96 well tissue culture plates
- EMEM
- 25 cm² tissue culture flask containing MDBK cells
- Trypsin-versene solution
- Pipettes
- Discard container
- FCS

Methods
- Clean the laminar hood with cotton/tissue paper soaked with 70% ethanol and sterilize the hood for about 30 min by putting on the UV light.
- Serial 10 fold dilutions of BHV-1 are made in EMEM containing 2% FCS in the following manner in the 24 well plate/tubes.
- Tube 1, 0.1 ml virus + 0.9 ml EMEM = 10^{-1}
  Tube 2, 0.1 ml of tube 1 + 0.9 ml EMEM = 10^{-2}
  Tube 3, 0.1 ml of tube 2 + 0.9 ml EMEM = 10^{-3}
  Tube 4, 0.1 ml of tube 4 + 0.9 ml EMEM = 10^{-4}
  Tube 5, 0.1 ml of tube 5 + 0.9 ml EMEM = 10^{-5}
  Tube 6, 0.1 ml of tube 6 + 0.9 ml EMEM = 10^{-6}

Tube 7, 0.1 ml of tube 7 + 0.9 ml EMEM = 10^{-7}
Tube 8, 0.1 ml of tube 8 + 0.9 ml EMEM = 10^{-8}
Tube 9, 0.1 ml of tube 9 + 0.9 ml EMEM = 10^{-9}
Tube 10, 0.1 ml of tube 10 + 0.9 ml EMEM = 10^{-10}

- Label the tubes/wells properly
- One 25 cm^2 flask containing the MDBK cells is taken and decan the medium in the discard container. Add one ml of trypsin-versene solution and rinse the cells properly and decant the trypsin solution. Repeat the trypsinization process once again.
- Keep the bottle in the incubator at 37°C for 2-3 minutes.
- When the cells start detaching from the plastic surface, add 3 ml of a EMEM containing 10% FCS and do gentle pipetting to make the uniform cell suspension having a concentration of 2X 10^5 cell/ ml with growth medium.
- 0.1 ml of each virus dilution are added to at least 5 wells of 96 well plate. Add 0.1 ml of cell suspension.
- Seal the plate and incubate at 37°C for 48 hours in the CO_2 incubator.
- Clean the laminar hood with cotton/tissue paper soaked with 70% ethanol and sterilize the hood for about 30 min by putting on the UV light.
- Reading is taken after 48 hours of incubation and recorded in a register.

Dilution	Infected/total	Cumulative infected (A)	Cumulative uninfected (B)	Ratio (A/B)	% infected
10^{-5}	5/5	9	0	9/9	100
10^{-6}	3/5	4	2	4/6	66.7
10^{-7}	1/5	1	6	1/7	14.3
10^{-8}	0/5	0	11	0/11	0

So, $TCID_{50}$ or 50% end point means the dilution that infect only 50% of the inoculated host system/cell wells and lies between 10^{-6} (66.7% infected) and 10^{-7} (14.3 % infected) dilutions.

The proportionate distance between these

$$P.D. = \frac{\% \text{ positive above } 50\% - 50}{\% \text{ positive above } 50\% - \% \text{ positive below } 50\%} = 0.3$$

So, 50% infectivity end point is there at a dilution $10^{-6.3}$

The titre of the original virus suspension = $10^{6.3}/0.1$ ml

= $10^{7.3}/1$ ml

21

COLLECTION, PRESERVATION & TRANSPORT OF VARIOUS SPECIMENS FOR LABORATORY DIAGNOSIS

The tentative diagnosis of viral diseases of livestock can be made by clinical symptoms alone. The clinical diagnosis is substantiated by the laboratory to obtain a precise, correct and prompt diagnosis. The success however depends on the collection of suitable materials in proper transport medium at a particular stage of disease condition and its dispatch to the laboratory in a manner that little or no alterations occurs in the specimen. This would facilitate the isolation and identification of the causative viral agents easily.

A detailed clinical antemortem and postmortem examination of animal is most important for the selection of material. Three different kinds of specimens are collected (a) specimens for virus isolation (b) specimens for serological examination and (c) specimens for direct examination.

Blood, serum, tissue samples, urine, faeces, milk and discharges should be collected when the concentration of virus/antibodies are highest in concentration. The titre of the virus is usually highest at the affected sites during early stages of the diseases. In diseases with viraemia, peak titre of virus often coincides with peak pyrexia and may precede the onset of clinical symptoms.

Post mortem examination should be carried out as soon as possible after death as putrefaction and autolysis takes place rapidly. Specimens for virus isolation are taken from the affected regions and have been given in the table.

Positive findings will include isolation and demonstration of virus or virus specific antigen and/or the demonstration of virus specific antibody. For this, clinician should supply adequate clinical details with each specimen, including age, species, breed, sex and date of onset of the disease with full clinical history. The amount of virus in an affected animal is usually maximum at or very shortly after the appearance of symptoms. There is generally a rapid decline after this. So, it is important to take any specimens for virus isolation as early as possible after the onset of the disease.

Suitable specimens : Generally, different specimens are required for virus isolation, direct examination and serology.

Type of specimens to be collected in different diseases

S. No.	System	For isolation	Direct examination	Serology
1.	Respiratory	Throat swab, nasopharyngeal secretion	Nasopharyngeal secretion	Paired sera
2.	C.N.S.	CSF, throat swab, faeces, blood	CSF, corneal impression smear	Paired sera

Contd...

Collection, Preservation & Transport

3.	Skin	Papular/vesicular/pustular fluid, crusts, nodules etc.	CSF, corneal impression smear	Paired sera
4.	Eye	Conjunctival swabs	CSF, corneal impression smear	Paired sera
5.	Liver	Blood	faeces	Paired sera

Post-mortem specimens : Specimens can also be taken from the animals after death.

Postmortem specimens to be collected in different diseases

S. No.	System	Specimens
1.	Respiratory	Lungs, tracheal swab
2.	CNS	Meninges, spinal cord, brain, CSF
3.	Cardiovascular	Heart, blood
4.	Skin disease	Papular/vesicular/pustular fluid, crusts, nodules etc.
5.	Liver	Liver, blood
6.	Abortion	Foetus, placenta

Collection of specimens : Specimens should be taken as early as possible in the acute phase of the illness. To check the bacterial contamination, all samples including blood samples should be taken aseptically. Swabs and other samples should be placed in virus transport medium (VTM). This is actually a buffered solution (PBS, HBSS, Earle's solution, NSS etc.) containing protein and antibiotics. Protein acts as a protective agent and to be added @ 1% in the form of BSA or skimmed milk powder or foetal calf serum (FCS). The source of protein is not important provided the protein is sterile, free from viruses and antibodies or viral inhibitors. Antibiotics usually penicillin and streptomycin but gentamycin and antifungal compounds may also be used to check the bacterial or fungal growth and should be included at concentration inhibitory to bacteria/fungi but nontoxic to cell culture. It must also be remembered that rickettsiae and chlamydial may be susceptible to penicillin, which should not be added to transport media for these agents. VTM should be prepared sterile and distributed in 2 ml quantities in suitable sterile glass/plastic containers. A pH indicator should be included and VTM should always maintain the neutral pH. The tissue specimen for histopathological examination should be collected and fixed as soon as after death to avoid autolytic changes. The tissue of appropriate size should be collected in 10% formol saline in a sterile container.

1) *Nasal swabs* : The cotton swabs are made and sterilized by autoclaving. Now-a-days ready made sterilized cotton swabs are available in the market. To take nasal swabs, the animal is kept in a standing position with head slightly raised. The swab is moistened with transport medium and gently introduced into the nostrils, rotated gently to be coated with nasal secretion. After removal, the swab is broken off into a tube or bottle or transport medium. Throat swabs are also collected in a similar manner.

2) *Rectal swabs* : The cotton swabs are made and sterilized by autoclaving. The swab is moistened with transport medium and gently inserted through the anal sphincter and gently rotated. The swab is finally withdrawn and broken off into the tube with transport medium.

3) *Faecal samples* : Although faeces are naturally contaminated with bacteria, it is essential that stools are collected aseptically. Diarrhoeic faeces should be collected directly from the animal rather than from the ground with the help of a cotton swab or gloved finger. Plastic bags may also be used.

4) *Vesicles and vesicular fluid* : The vesicles on the mucous membranes such as tongue, gum, dental pads are

collected using sterile precautions. In large animals, the animal is restrained, the tongue is taken out using muslin cloth/surgical gauze and vesicles are collected using sterile forceps/scissors in the transport medium. Vesicular fluid also can be collected with the help of cotton swabs and broken off into a bottle of transport medium.

5) *Cerebrospinal fluid (CSF)* : The CSF can be collected from the cerebellomedullary cistern or the lumbosacral space. The lumbar puncture is the preferred one. A lumbar puncture is made between 1st and 2nd lumbar vertebrae. A needle of 1.5 cm is inserted between the vertebrae and CSF withdrawn with a syringe.

6) *Lesion scrapings* : In cutaneous lesion, the skin is first sterilized by swabbing with diethyl ether. A sterile scalpel is used to remove suitable material and put in the transport medium for further laboratory examination.

7) *Urine* : The urethra is washed and sterilized and the urine is collected directly into a sterile container.

8) *Saliva* : It is directly collected into a sterile container.

9) *Blood* : The skin is first sterilized by swabbing with ethyl alcohol or tincture of iodine. The needle is carefully inserted into the vein with a 10 ml syringe fitted with appropriate needle required for different species of animals and requisite volume of blood is collected in a sterile container containing the anti-coagulant. A sterile cotton is held to arrest the bleeding. Blood is collected from jugular vein of sheep, goat, equines and bovines, from anterior vena cava of pigs and from tarsal veins of dogs.

10) *Serum* : The blood is collected as described above in sterile glass test tube and kept in flat position. After clotting, liquid phase is collected as serum for serological examination. It is necessary to collect two specimens at an interval of 2-3 weeks.

22

COLLECTION OF MATERIALS FOR DIAGNOSIS OF VIRAL DISEASES

Bovine

1.	PPR	Eye, mouth, rectal swab in PBS on ice, 10 ml blood in anti-coagulant at height of temperature, Prescapular lymph node, spleen, piece of intestine on ice, lung, liver, spleen, tonsil in 10% neutral formol saline solution.
2.	RP	Same as PPR
3.	FMD	Vesicular fluid in transport medium (TM) on ice, vesicular epithelium i.e. tongue, gum, dental pad lesions in 50% phosphate buffered saline on ice. 10 ml blood in anti-coagulantat at height of temperature, pieces of heart and other organs from calves in 10% neutral formol saline solution.
4.	IBR/IPV, Bovine mamilitis,	Nasal, conjunctival and genital swabs and foetal tissues in transport medium, paired serum samples on ice, affected trachea, turbinate bones,

Contd...

	Parainfluenza 3, Adenovirus	foetal tissues etc in 10% formol saline and in transport medium (TM). From bulls semen and preputial washing in TM and paired serum on ice.
5.	BVD	Blood in EDTA, paired serum samples, semen, intestine, lymph node and spleen on ice, pieces of brain, liver, spleen, kidney, adrenals and nasal mucosa in 10% formalin.
6.	Ephemeral fever	Blood in EDTA on ice and paired serum samples.
7.	Malignant catarrhal fever	Blood in EDTA, paired serum samples, liver, spleen and lymph node on ice. Pieces of brain, liver, spleen, kidney, adrenals and nasal mucosa in 10% formalin.
8.	Cow pox/ buffalo pox	Scabs in sterile containers on ice, scabs in 50% buffered glycerin and skin lesions on 10% neutral formol saline separately.
9.	Enzootic bovine leukosis	Blood in EDTA, tumour tissues, lymph node, abomasums, heart, spleen, intestine, liver, lung and uterus.
10.	Rabies	Similar to canines

Canines

1.	Rabies	Half of brain, salivary gland in 50% phosphate buffered glycerin on ice in leak proof hard box and the rest of the brain in 10% neutral formol saline solution. Small pieces of hippocampus and brain (cerebellum, medulla, cerebrum, spinal cord) in 50% buffered glycerin and 10% neutral formol solution separately duly sealed and packed in thick poly bag and hard box labeled "suspected for rabies". All staff attending post-mortem be vaccinated against rabies.
2.	ICH	Lung, liver, gall bladder, spleen and kidney in TM on ice and 10% formol saline.

Contd...

3.	CD	Pieces of lung, urinary bladder, liver, trachea, stomach wall, brain etc in 10% formol saline, spleen and liver in TM on ice.
4.	CPV	Pieces of liver, spleen, intestines, stomachs and lymph nodes in 50% buffered glycerin saline and internal organs in neutral formol saline solution.

Ovine and caprine

1.	Sheep pox and goat pox	As described in bovines.
2.	Rabies	As described in canines.
3.	Bluetongue	Blood at the height of body temperature in heparin (10 units/ml) or EDTA, paired serum samples on ice. Spleen and lymph nodes (5-10 g) on ice. Pieces of intestine, lymph nodes and internal organs etc in 10 % neutral formol saline solution.
4.	Rift valley fever	Blood in heparin or EDTA (5 ml), liver (5g), spleen, brain, aborted foetus on ice, liver, spleen, lungs and lymph nodes in 10% formol saline for histopathology.
5.	CAE/ Maedi/ Visna	Citrated blood, CSF and saliva, lungs, spleen and brain in TM on ice and brain lungs and spleen in 10% neutral formol saline.

Equine

1.	Equine influenza	Nasal swabs in PBS or HBSS on ice and paired serum samples.
2.	Equine infectious anaemia	Blood and serum on ice and all internal organs in 10% neutral formol saline solution
3.	African horse sickness	Blood in EDTA, paired serum samples, spleen, brain and lung in 50% buffered glycerin and 10% formol saline separately.

Contd...

4.	Equine herpes virus, Equine rhinopneumonitis	Nasal swabs, liver, lung, spleen and thymus from aborted foetus and paired serum samples on ice and blood in EDTA.
5.	Equine encephalomyelitis	Citrated blood, paired serum samples and brain on ice.

Porcine

1	Swine fever	Heparinized blood at the height of temperature in sterile vials or test tubes and serum on ice from live animals. Spleen, lymph node and pancreas in 50% buffered glycerin saline, pieces of brain, lung, intestines and kidney in 10% neutral formol saline solution.
2.	African swine fever	Blood in heparin or EDTA, spleen, tonsil, kidney, lymph node on ice and paired serum samples,
3.	TGE	Faeces, small intestine, lung and udder on ice and paired serum samples.
4.	PRRS	Lung, liver, lymph node, tonsil, spleen, heart, brain, ascitic fluid and paired serum samples on ice.
5.	Rabies	As in canines.
6.	Aujesky's disease	Brain, spinal cord, spleen, kidney and lymph node in TM on ice and in 10 neutral formol saline. Blood in heparin/EDTA and paired serum samples.
7.	Swine influenza	Nasal swabs, lungs and trachea in TM on ice.

Records : While sending the materials for laboratory examination, disease conditions and post-mortem findings should be sent. In case of outbreaks, their origin, spread, nature of the disease, age, sex, breed, species, attack rate, death and survivals, vaccination status of animals etc. should be mentioned. The kind of medicine used to treat the animals should also be mentioned. The specimens must be clearly labelled and examination required and disease suspected should be mentioned. The disease suspected will indicate the material to be collected and submitted for laboratory examination.

Packaging : After collecting the specimens in VTM, it is to be packed properly before sending it to a suitable laboratory. Packaging has three main purposes : to maintain the specimen's viability, to prevent it leaking outside the package and to prevent cross contamination. The packaging should be in three layers.

1) *A primary receptacle* : containing the specimen itself. This must be water tight and where volatile buffers are used it must be air tight. Plastic or glass vials/containers with screw caps are usually used as primary receptacle. Screw caps and other lids should be taped to prevent accidental loosening.

2) *Secondary packaging* : This is a water tight secondary layer enclosing enough absorptive material (tissue paper, absorbent cotton, wool etc.) to absorb all the fluid of the specimen in case of leakage. Polythene bag is generally used as secondary packaging.

3) *Outer packaging* : It is intended to protect the secondary packaging from outside influence viz., physical damage and water during transit. Metal, wooden or plastic container/box with tightly fitting lid are generally used as outer packaging.

Refrigeration : Biological materials survive better at low temperature. So, it is desirable to maintain the low temperature during the shipment of the specimens. Refrigeration may be achieved by using solid CO_2, wet ice or frozen pads. Repeated freezing and thawing should be avoided as this may cause loss of titre. The outer package should be kept inside an icebox containing refrigerant in such a way as to keep the two in contact.

Transport : Collected materials/samples in VTM can be sent to a nearby laboratory for diagnosis by keeping it on ice through a messenger. However, samples should be properly packed before sending it to a distant laboratory through messenger, mail, post, freight, courier etc. Viruses require living cells to grow. The amount of virus in a specimen will not increase after it has been taken but will decline and the rate of declining depend on the temperature and other condition. It is therefore important that the time in transit should be as short as possible. The temperature is the major factor in virus survival during transit. Viruses vary considerably in heat stability. Surface proteins are denatured within few minutes at temperature of 55-60⁰C with the result that the virion is non-infectious, because it is no longer capable of normal cellular attachment and/or uncoating. At ambient temperature, the rate of decay of infectivity is slower but significant, especially in hot summer weather or in the tropics in any season. Viruses must be stored at low temperature, 4⁰C (ice or a refrigerator) is usually satisfactory for a day or so but long term preservation requires temperature well below zero. Two convenient temperatures are -79⁰C, the temperature of frozen CO_2 (dry ice) and of some freezers or -196⁰C, the temperature of liquid N_2. As a thumb rule, the half life of most of the viruses can be measured in seconds at 60⁰C , minutes at 37⁰C, hours at 20⁰C , days at 4⁰C and years at -79⁰C or lowers.

23

PROCESSING OF LABORATORY SPECIMENS, INOCULATION INTO LABORATORY ANIMALS AND EMBRYONATED EGGS FOR PATHOGEN ISOLATION

Processing

On arrival in the laboratory, the specimens are processed immediately or refrigerated until processed. For inoculation into animal/cell culture/embryonated eggs (EE), swabs are shaken in fluid medium, faeces are dispersed in fluid medium and tissue specimens are homogenized in a high speed blender/pestle and mortar with the help of an abrasive like powdered glass or sand. Good quality hard sand should be properly washed in distilled water, treated with HCl, washed thoroughly, dried and sterilized before use. Generally the infective tissue is homogenized in a few ml of buffer usually having a pH of 7.2 to 7.3 to make about 10% tissue suspension. The suspension is centrifuged at about 1000g for 10-15 minutes to sediment the tissue debris and coarse materials. In the supernatant, antibiotics are added @ 10,000 units of penicillin and 10 mg of streptomycin per ml of fluid and allowed to act for about 30 minutes at room temperature. The supernatant is collected and usually passed through a 0.45 µM disposable syringe filter to remove contaminating non-viral organisms. This inoculum is now ready to be inoculated into cell culture/animal/EE. Clinical specimens processed in this way are generally suitable for detection of viral antigens/DNA/RNA by *in vitro* tests.

Inoculation into laboratory animals : In most of the cases, attempts are made to cultivate the viruses in embryonated eggs or cell cultures which are more convenient and economical than animals. However, to find out the course of the disease, the clinical symptoms, pathogenicity and P.M. lesions, the viruses have to be inoculated in the natural hosts and laboratory animals. The inoculation of specimens in laboratory animals are preferred over the natural hosts because they are cheap, readily available and easily restrained. But there are some viruses which can be cultivated only in natural hosts. The animals used to inoculate the specimens should be healthy, disease free and free from antibodies to pathogens for which the investigation is carried out.

Some viruses have preference to grow in certain tissues like skin, respiratory system, nervous system, CNS etc while others can grow in all organs and tissues of the animals. The route of inoculation has to be selected according to the possible tissue predilection of the viruses. Different routes of inoculation may be I/M, S/C, I/V, I/P, I/N and I/C and the laboratory animals for virological studies are mice, guinea pigs, rats, rabbits, hamsters, chickens etc.

Embryonated eggs inoculation : Embryonated eggs form a good medium to cultivate viruses particularly the viruses of poultry, being a natural host, free from antibodies, small to handle and naturally protected from contamination. The eggs used for this work should preferably be from specific pathogen free (SPF) flock or at least disease free flock because some of the organisms get passed into eggs from the layer hens. The eggs for this work should be kept in a small egg incubator of about 100 or more eggs capacity kept at 37°C. Humidity is maintained at 60% by keeping a trough containing water. Eggs should be turned once or twice a day.

Depending on the type of virus, different routes of inoculation are used.

(A) *Inoculation into Yolk sac* : It is done in 5 to 7 or 8 days old fertile eggs to grow rabies, avian encephalomyelitis virus, Marek's disease virus, chicken infectious anaemia virus, infectious bronchitis virus, chlamydia and mycoplasma.

Method : The fertile eggs are candled in a dark room in an egg candler which can be made by making a hole in an ordinary wooden box with an electric bulb inside the box. Discard eggs which are cracked, infertile and having dead embryos. Living embryos will show blood vessels of CAM and dead ones will not show blood vessels. By candling, mark the air space with a lead pencil. At the top of the air space make a small hole after disinfecting it with 70% ethanol with the help of dentist's drill. The specimen is taken in a small tuberculin (1 ml) syringe fitted with 23 gauge and 1.5" long needle. The egg with hole is kept vertically in an egg tray keeping the hole top most. The needle is introduced vertically through the hole till almost entire length of the needle enters the egg. Now about 0.2 to 0.25 ml of inoculum is slowly pushed from the syringe into the egg. The hole is sealed immediately with hot wax. The eggs are incubated at 35-37°C under 60-70% humidity. They are kept vertically, undisturbed for 10 days before harvesting the embryos. For this, eggs are candled, cut the egg shell 1 mm above the air space mark. The shell cap is removed with a scalpel. Now, with the help of a sterile forceps CAM and amniotic sac are removed from the top and embryos and yolk sacs are harvested and noted for any changes.

(B) *Inoculation into CAM* : This method is generally used for isolation of IBD, avian pox, cow pox, sheep pox, goat pox and buffalo pox viruses.

Method : Embryonated eggs incubated for 10-12 days are taken from disease free and preferably SPF flock. The eggs are candled and one triangle of 1-1.5 cm sides is marked with a pencil. The area should not have any major blood vessels. The dentist's drill is used to carefully cut the shell along the lines of the triangle. The CAM lies just below the shell and care must be taken not to damage it. The cut piece of shell is slowly and carefully lifted with a sharp scalpel or needle. With a sterile needle, a small slit is made in the shell membrane. Now a small hole is drilled through the shell and shell membrane in the centre of the air cell. A rubber bulb is used to apply suction at the hole. By suction, an artificial air cell forms in the area of the cut triangular shell because air can enter through the slit made in the shell membrane, in the triangular area. The outer wall of the air cell is formed by the shell membrane and the inner wall by the CAM. 0.1 ml to 0.5 ml inoculum can be put on the CAM, by a syringe, through the slit. After inoculation, the openings on the natural air cell and the artificial air cells, are sealed with the molten paraffin wax.

(C) *Inoculation into allantoic cavity* : This method is generally done to isolate ND, avian influenza and adenoviruses.

Method : 9 to 11 days old embryo eggs are used. The eggs are candled and air sac is marked with pencil. A small hole is drilled about 3 mm inside the margin of the air cell, piercing shell and shell membrane. The egg is kept vertically with the air cell on top. A tuberculin syringe fitted with a 5/8" long needle is filled with the inoculum. The needle is pushed vertically through the hole and 0.3 ml of inoculum is pushed into the allantoic sac. The hole is sealed and incubation is done as above.

Processing of Laboratory Specimens | 131

(D) *Intravenous inoculation* : This method is generally used to inoculate the bluetongue virus. This method is extremely difficult and require special expertise and skill. 9 to 12 days old chicken embryo are candled and superficial large blood vessel is marked with a pencil. The egg surface is disinfected and a triangle is cut out around the marked area of blood vessel leaving shell membrane intact. The shell is removed and eggs are kept on the tray with the exposed vein on the top. A drop of mineral oil is applied on the triangular area to make the vein more distinct. A tuberculin syringe fitted with 26 gauge needle is pierced slowly parallel to the vein and 0.2 ml of inoculum is injected followed by incubation at 37^0C.

S. No.	Route of inoculation	Age of embryo (days)	Volume of inoculation (ml)	Type of virus
1.	CAM	10-12	0.1	CD, ILT, fowlpox, buffalo pox, sheep pox, goat pox, PRV, HSV, VZV, Vaccinia, IBD, reovirus, ALC
2.	Allantoic cavity	9-12	0.1	RD, IB, avian influenza, mumps
3.	Amniotic cavity	7-15	0.1	Influenza, mumps, duck viral hepatitis
4.	Yolk sac	5-8	0.1	Mumps, rabies, IBH-hydropericardium syndrome, chicken infectious anaemia, MD, avian encephalomyelitis, ALC
5.	I/V	9-12	0.1	Bluetongue. ALC
6.	I/C	8-14	0.1	Rabies, HSV

Points to be remembered while inoculating embryonated eggs

(1) Eggs from hens vaccinated against the disease under investigation should not be used.

(2) Eggs must be sterilized with cotton soaked with ethyl alcohol before inoculation.

(3) Fertile eggs with well developed blood vessels and jerking movements indicating a developing embryo should be used.

(4) The embryo should be examined immediately after death preferably after keeping at 4^0C for ½ to 1 hour to minimize bleeding.

(5) Eggs should be opened near a flame or inside a laminar flow if the material is to be used for further tests or passage.

(6) Allantoic fluid is removed by breaking the egg shell on the air sac, then removing the shell membrane and CAM and fluid is sucked with a pipette.

(7) CAM should be examined by keeping the membrane in a petridish against a black background.

(8) About 15% embryos die normally and should not be regarded as death due to virus.

(9) Some viruses need blind passages of embryonic material to detect pathogenicity in embryo.

24

SERUM NEUTRALIZATION TEST

Principle

The infectivity of a virus may be neutralized by specific antibody when they are mixed and incubated. In some cases, the addition of homologous or guinea pig complement to the mixture enhances neutralization, particularly when early antiserum is used. Serum must be inactivated by heating at 56^0C for 30 minutes to inactivate complement or to remove non-specific virus inhibitors. Serum-virus mixtures are inoculated into appropriate cell cultures, which are then incubated until the only virus controls develop cytopathic effects. Antibody by neutralizing the infectivity of the virus, protects the cells against virus destruction.

Now-a-days, most neutralization tests are now conducted in disposable non-toxic sterile 96 wells plastic flat-bottomed plates in which a cell monolayer is established. Virus –antiserum mixtures can be added to established monolayers, or more usually, serum dilutions are made in the wells, a standard amount of virus is added, and the mixture incubated, after which cells are added. In the standard neutralization test the end point is read by cytopathic effect, the titer of the serum being defined as the highest dilution that inhibits the cytopathic effect. The SNT can be performed in 2 ways. In the constant-serum diluted virus (α procedure), a constant concentration of antibody is added to an equal volume of serial tenfold dilution of virus. In the constant virus diluted antibody (β procedure), a constant concentration of virus, usually 100 to 200 $TCID_{50}$ is added to an equal volume of serial two fold dilutions of antibody. Although the α and β procedures of the SNT are specific, they do not give results directly. Usually, the α procedure is used to identify a virus and the β procedure to measure the antibody titer. In both the procedures, after an incubation of 1 to 2 hours at 37°C, the mixtures are inoculated into a suitable host system.

Sample No		1	2	3	4	5	6	7	8	9	10	11	12	Sample No
1	A													9
2	B													10
3	C													11
4	D													12
5	E													Known positive
6	F													Known negative
7	G													Virus Control
8	H													Cell control

β Procedure

1) Make two fold dilutions of the test serum samples (only two dilutions, say 1:2 and 1:4) along with the known positive and known negative serum samples and add 200 μl / well of 24 well plate or sterile test tube inactivating all the antisera samples by heating at 56°C for 30 minutes in water bath. Dilution of the serum samples should be made in cell culture maintenance media containing 2% faetal calf serum.

2) Make 100 $TCID_{50}$/ 200 μl virus suspension in the maintenance media containing 2% faetal calf serum.

3) Add equal quantities of virus suspension (200 μl) per well or test tube containing different dilutions of serum samples. Mix the serum and virus mixture properly and incubate for 2 hours at 37°C.

4) Add each dilution of serum sample containing virus suspension in triplicate wells (three wells for each dilution) @ 100 μl /well in the 96 well plate. (Ex Sample no. 1, 1:2 dilution A1, A2 and A3 and 1:4 dilution A4, A5 and A6. Similarly add serum sample 2, from B1 to B6 and so on. Similarly add the known positive (1:2 dilution in E7 to E9 and 1:4 dilution in E10 to E12) and known negative (1:2 dilution in F7 to F9 and 1:4 dilution in F10 to F12) serum samples. The final dilution of serum will be 1:4 and 1:8 after mixing with equal quantities of virus suspension.

5) Add the suitable cell (BHK21 for bluetongue virus and MDBK for IBR virus) at the concentration of $2X10^4$ cells/ 100 μl/per well. Incubate the micro-SNT plate in the carbon dioxide incubator for 48 hours. It means all the wells will contain 100 μl of cells having a concentration of $2X10^4$ cells/100 μl.

6) Note down the reading and express the titre of the serum. If a particular serum sample contains the specific antibody, it will neutralize the virus and there will be no

CPE and confluent monolayer will be visualized. If the serum sample does not contain the specific antibody, it will not be able to neutralize the virus and CPE will be visualized due to action of the virus on the susceptible cells. The titre of the serum is expressed as dilution of serum which could able to neutralize the 100 $TCID_{50}$ virus. If the dilution of 1:4 of a particular serum is able to neutralize the virus but not the 1:8 dilution of the serum, then titre of the serum will be 1:4.

Note

1) In the cell control wells, there will be 100 µl maintenance media containing 2% FCS and 100 µl cells.
2) In the virus control wells, there will be 50 µl virus, 50 µl maintenance media and 100 µl cells.
3) The total volume of all the wells of the plate will be 200 µl.
4) Swine fever virus does not produce any CPE in PK-15 cell lines. So, indirect immunofluorescent technique should be carried out taking the content of all the wells to find out the positive and negative immunofluorescence and accordingly titre of the serum should be expressed.

Determination of SN titre by alpha (a) procedure (Quantal method)

Virus dilution	Control (Negative serum or diluent + virus)		Test (Test serum + virus)	
	Infected	Noninfected	Infected	Noninfected
10^{-3}	5	0	5	0
10^{-4}	5	0	3	2
10^{-5}	5	0	1	4
10^{-6}	3	2	0	5
10^{-7}	1	4	0	5

Serum Neutralization Test

Virus dilution	Control Negative serum or diluent +virus		Accumulative Value		Ratio	% Infected
	Infected	Noninfected	Infected	Noninfected		
10^{-3}	5	0	18	0	18/18	100
10^{-4}	5	0	13	0	13/13	100
10^{-5}	5	0	8	0	8/8	100
10^{-6}	3	2	4	2	4/6	67
10^{-7}	1	4	1	6	1/7	14

Virus dilution	Test Test serum +virus		Accumulative value		Ratio	% Infected
	Infected	Noninfected	Infected	Noninfected		
10^{-3}	5	0	9	0	9/9	100
10^{-4}	3	2	4	2	4/6	67
10^{-5}	1	4	1	6	1/7	14
10^{-6}	0	5	0	11	0/11	0
10^{-7}	0	5	0	16	0/16	0

Calculation by reed and muench method

The proportion distance (PD) between the 2 dilutions (10^{-6} and 10^{-7}) of control or negative serum :

$$= \frac{\text{\% infection above 50\% -50}}{\text{\% infection above 50\% - \% infection below 50\%}} = \frac{67-50}{67-14} = \frac{17}{53} = 0.3$$

Titer of the control virus = $10^{6.3}$ TCID$_{50}$

The proportion distance (PD) between the 2 dilutions (10^{-4} and 10^{-5}) of test serum :

$$= \frac{\text{\% infection above 50\% -50}}{\text{\% infection above 50\% - \% infection below 50\%}} = \frac{67-50}{67-14} = \frac{17}{53} = 0.3$$

Titer of virus with serum = $10^{4.3}$ TCID$_{50}$

So, the neutralization index is = $10^{6.3} - 10^{4.3} = 2$.

Determination of SN titer by beta (β) procedure (Quantal method)

Serum dilution	Test results		Accumulative		Ratio of protected	% protected
	Protected	Infected	Protected	Infected		
1:8 ($10^{-0.9}$)	5	0	11	0	11/11	100
1:16 ($10^{-1.2}$)	4	1	6	1	6/7	86
1:32 ($10^{-1.5}$)	2	3	2	4	2/6	33
1:64 ($10^{-1.8}$)	0	5	0	9	0/9	0

Proportionate distance : 0.7

Exponential of dilution which provided exactly 50% protection = P.D.x[E.D. next below 50%-E.D. next above 50%] + E.D. next above 50% = -1.4

So, the 50% neutralization end point = $10^{-1.4}$ or 1:25

25

ENZYME LINKED IMMUNOSORBENT ASSAY (ELISA)

Early and accurate diagnosis of disease is of prime importance for both the successful treatment of the patient and in limiting the spread of disease in the community. It is in this context that immunodiagnostic methods are now being used extensively since they can be extremely sensitive and highly specific. Among the most successful methods are those utilizing labelled antibodies or antigens. Examples of these are immunofluorescence in which antibodies labelled with fluorescent dyes are utilized and radioimmunoassay in which antigens or antibodies are labeled with isotopes. Both of these procedures have proved to be of great value for particular purposes. For example, immunofluorescence is used for identification of viruses in clinical samples and radioimmunoassay is used for determination of the extremely low levels of hormones in body fluids. Fluorescent antibody methods are time consuming and require costly UV microscope and radioimmunoassay is expensive and employs reagent which have a short shelf life and are potentially hazardous. So, the researchers have developed the enzymes as markers, in enzyme immunoassay, devoid of these problems.

Enzyme linked immunosorbent assay (ELISA) is a non-isotopic immunoassay in which the antigen and antibody complex is detected by using specific enzyme antibody conjugate and substrate. The description of ELISA was first given by Engvall and Perlmann in 1971 and the technique was first applied in the diagnosis of viral diseases by Voller and Bidwell in 1976. ELISA is simple, sensitive, specific and precise and is occupying the important places in serological diagnosis of a number disease caused by various pathogens. The test can be used to identify the causative agents of infectious (virus, bacteria etc) diseases, detect the presence of specific antibody, quantify the antigens and antibody etc. It can also be used in the diagnosis of diseases caused by protozoa, rickettsia, mycoplasma, fungi etc and the estimation of different hormones in the biological fluid. Additional advantage of this assay is that a large number of samples can be handled at a time.

Principle : The basic principle of ELISA is the detection of solid phase bound antigen and antibody complex by a color reaction using specific enzyme and substrate. An antigen or antibody which is usually protein in nature can be adsorbed passively on solid phase. If an unknown antigen is adsorbed on a solid phase, its presence can be identified by adding specific antibody covalently linked to an enzyme. If the unknown antigen is specific to the antibody then an antigen+ antibody+ enzyme complex will form and remain bind to the solid phase. To this bound complex if a substrate specific to the labeled enzyme is added then the enzyme will cause breakdown of the substrate resulting the colour development. This colour development can be detected visually or read in spectrophotometer (ELISA reader).

Classification : Various ELISA techniques developed can be classified into the following types.

1. Direct ELISA
2. Indirect ELISA
3. Sandwich ELISA/Immunocapture ELISA
4. Competition ELISA
5. Liquid phase blocking sandwich ELISA
6. Liquid phase blocking competition ELISA

Type of reagents

Solid phase : Though various solid phase such as paper, nylon, polystyrene beads, agarose, sephadex, 96 well microtitre plates and tube etc. can be used for ELISA, but the commonly used solid phase in different ELISA is 96 well microplate. Microplates for ELISA are of special type and manufactured from polyvinyl chloride or polystyrene. The ordinary 96 well plate for microcomplement fixation test, tissue culture or haemagglutination test should not be used in ELISA.

Microtitre pipettes : These are required to transfer accurate volume of different reagent in ELISA. Commercially available single channel and multichannel pipettes of different capacity are commonly used for this purpose.

Disposable tips : Tips made of plastic or polystyrene are commercially available. These tips can be fitted to the different multichannel and single channel pipettes and used for transferring the reagent in ELISA. Tips can be reused if washed properly, but the tips which are used for transferring conjugates should not be reused.

Washing solution : Careful washing of the ELISA plate at different steps is very essential. Commonly used washing solution and its composition is given in appendix. Washing of the ELISA plates can be carried out manually or by the use of machine (ELISA plate washer).

Coating buffer : The carbonate/bicarbonate buffer pH 9.6 is commonly used as coating buffer. The 0.1 M phosphate buffer saline (PBS) pH 7.2-7.6 can also be used as coating buffer, but the first one is found to be better.

Blocking buffer : Blocking buffer is used to prevent adsorption of nonspecific protein to the ELISA plate wells. To prevent this, certain immunologically inert substances are used in a dilution buffer. Any one of the these [Bovine serum albumin (BSA) fraction V (1-5%), FCS (1-5%), horse serum (1-5%), lactalbumin hydrolysate (1-5%) and skimmed milk powder (SMP) (3-5%)] can be used at a concentration shown against them in PBS as blocking buffer.

Enzymes : Different enzymes such as horse radish peroxidase (HRPO), alkaline phosphatase, β galactosidase, glucose oxidase and urease have been used by different workers in ELISA systems. However, HRPO is the enzyme of choice because it gives satisfactory and reproducible results and available commercially from different commercial companies.

Substrate solution : The choice of a substrate solution depends upon the type of enzyme. For one type of enzyme particular substrate should be used. The enzyme substrate combination is shown below.

Type of enzyme	Type of substrate
Horse Radish Peroxidase	Ortho-phenylenediamine(OPD)/ODD/ ABTS /TMB/5 amino salicylic acid and H_2O_2
Alkaline phosphatase	p-nitrophenyl phosphate (PNP)
β galactosidase	O-nitrophenyl-D-galactoside (ONPG)

OPD solution can be prepared by dissolving 40 mg of OPD powder in 100 ml of 0.1 M sodium citrate buffer (pH 5.0). Always prepare the solution just before use. OPD substrate solution is added in the test system using H_2O_2 @ 10 µl per 10 ml of OPD solution. H_2O_2 is to be added to the OPD solution immediately just before its use in the test.

Stopping reagent : The commonly used stopping reagent for OPD+ H_2O_2 is 1(M) H_2SO_4. This is prepared by adding 54 ml of concentrated sulphuric acid to 946 ml of distilled water. Solution can be stored at room temperature. Always add acid to the water not vice versa.

Direct ELISA : This assay can be used to detect unknown antigen. If antigen is unknown, then known specific antibody should be used for labeling with enzyme.

1) Coat the ELISA plate with test antigen by incubating overnight at 4°C or at 37°C for 1 hour.

2) Wash the plate with washing solution at least three time.

3) Add enzyme labeled antibody solution in the blocking buffer and incubate at 37°C for 1 hour.

4) Wash the plate with washing solution at least three time.

5) Add substrate solution and keep the plate in the dark for 10-15 min. Stop the reaction by adding stopping reagent and look for colour development visually or read in an ELISA reader.

Interpretation : Colour development indicates positive reaction and no colour indicates the negative reaction. Intensity of the colour is proportionate to the amount of antigen or antibody complex on the solid phase.

Advantage : Very simple and less time consuming.

Disadvantage : Non specific reaction if crude antigen is used for coating. There is a need to procure/prepare conjugate against all the pathogens. The conjugates are very expensive.

Note : Now-a-days, direct ELISA is not commonly used in any diagnostic laboratory.

Indirect ELISA : This assay is used to detect the unknown antibody against a particular pathogen.

1) Coat the ELISA plate with purified known antigen diluted in coating buffer and incubate at 4°C overnight or at 37°C for 1 hour.
2) Wash the plate with washing buffer for at least 3 times.
3) Add serum samples/antibody in blocking buffer and incubate at 37°C for 1 hour.
4) Wash the plate with washing buffer for at least 3 times.
5) Add enzyme labelled anti-species antibody (conjugate) in blocking buffer and incubate at 37°C for 1 hour.
6) Wash the plate with washing buffer for at least 3 times.
7) Add substrate solution and keep the plate in the dark for 10-15 min. Stop the reaction by adding stopping reagent and look for colour development visually or read in ELISA reader.

Advantage : More specific for detection and quantification of unknown antibody in serum samples.

Disadvantage : More time consuming and require anti species antibody. Antispecies antibody is prepared against the normal globulin of species from which serum samples are tested. For examples if the serum samples are tested from sheep then anti species antibody is prepared in species other than sheep by using normal ovine globulin.

Note : It is commonly used to detect the antibodies to BHV-1, bluetongue virus, PPR virus, FMD virus etc.

Sandwich ELISA : This assay is also called as capture or trapping assay, because it is used to capture or trapping antigen by specific antibody passively adsorbed on the solid phase.

Certain antigen can not be used to coat ELISA plates directly because of their very low concentration or due to the presence of high concentration of contaminating protein or any other unknown antigen from crude mixture. Such antigen can be detected by this assay.

1) Coat ELISA plate with specific antibody (suppose rabbit antibody against goat pox virus) solution diluted in coating buffer and incubate at 4°C overnight or at 37°C for 1 hour.
2) Wash the plate with washing buffer for at least 3 times.
3) Add antigen (test sample suspected for goat pox virus) solution diluted in blocking buffer and incubate at 37°C for 1 hour.
4) Wash the plate with washing buffer for at least 3 times.
5) Add antibody (guinea pig antibody against goat pox virus) directed against antigen. Incubate at 37°C for 1 hour.
6) Wash the plate with washing buffer for at least 3 times.
7) Add conjugate (anti-guinea pig globulin coupled with enzyme) and incubate at 37°C for 1 hour.
8) Wash the plate with washing buffer for at least 3 times.
9) Add substrate solution and observe for 10-15 minutes. Stop the reaction with stopping reagent and read.

Advantage : Specific for detection of unknown antigen present in very low level or in crude mixture.

Disadvantage : Antibodies against the antigen preferably raised in two species are required.

Note : The sandwich ELISA is used commonly to detect the PPR virus in the clinical samples. It is also used to detect and type the FMD virus present in the clinical samples and called typing ELISA.

Competition ELISA (direct antibody competition)

Apply to identify and quantify unknown antibody and to compare the relative affinity of binding of antibodies for the same antigen.

1) Coat the ELISA plate with purified antigen after diluting in coating buffer. Incubate at 4°C overnight or at 37°C for 1 hour.
2) Wash the plate with washing buffer for at least 3 times.
3) Add the test serum sample against the antigen followed by addition of the monoclonal antibody directed against the antigen and incubate at 37°C for 1 hour.
4) Wash the plate with washing buffer for at least 3 times.
5) Add enzyme labeled anti-mouse antibody (conjugate) and incubate at 37°C for 1 hour.
6) Wash the plate with washing buffer for at least 3 times.
7) Add substrate solution and observe for 10-15 minutes and stop the reaction and read.

Interpretation

1) No colour development indicate the specificity of the test antibody and complete saturation of binding sites of the antigen by the antibody leaving no binding site for the monoclonal antibody.
2) Less colour indicate low concentration of the antibody leaving binding site for the monoclonal antibody.
3) Intense colour development indicate that the suspected antibody is not specific for the antigen leaving all binding sites for the monoclonal antibody.

Note : It is commonly used for detection of antibodies to PPR virus, BHV-1, bluetongue etc.

Direct ELISA (for detection of antigen)

Test antigen + Antibody against a particular antigen coupled with enzyme (conjugate) + substrate = Colour or no colour.

Colour : Positive for that particular antigen.

No colour : Negative for that particular antigen.

Indirect ELISA (for detection of antibody)

Known antigen + test antibody + anti-species antibody coupled with enzyme (conjugate) + substrate = Colour or no colour.

Colour : Positive for that particular antibody in the test serum sample.

No colour : Negative for that particular antibody in the test serum sample.

Sandwich ELISA (for detection of antigen)

Antibody against the known antigen + test antigen + another antibody against the known antigen + Anti-species antibodies coupled with enzyme (conjugate) + substrate = Colour or no colour.

Colour : Positive for that particular antigen.

No colour : Negative for that particular antigen.

Competition ELISA (for detection of antibody)

Known antigen + test antibody and monoclonal antibody against the antigen + anti-mouse immunoglobulin coupled with enzyme (conjugate) + substrate = Colour or no colour.

Colour : Negative for the particular antibody in the test serum sample.

No colour : Positive for the particular antibody in the test serum sample.

26
BIOSAFETY MEASURES AND RISK MANAGEMENT

Precautions to be taken while working in the laboratory for personal safety and to prevent any hazards.

1. Be vaccinated against dangerous viruses *viz.*, rabies and hepatitis B as persons can contract infection while working with these organisms.

2. Use eye protection viz., goggles, spectacles etc to avoid splashing microbes into the eyes.

3. Do not rub eyes with contaminated hands or objects.

4. Do not eat or drink anything in the microbiology laboratory as it will contaminate the laboratory.

5. Do not create aerosols of microbes as one may be infected through inhalation.

6. Carefully dispose of the cultures of pathogenic organisms by using disinfectants or autoclaving.

7. Keep microbes away from open wounds.
8. Always wear apron, face mask, gloves and cap while handling any organisms.
9. Alcohol lamp can spill and create accidents and special precautions to be taken while using it.
10. Handle the blood and body fluids as they may carry dreadful pathogens.
11. Carefully handle the sharp objects as you may sustain injury from it.
12. Wash the hands after work with soaps and water.
13. Avoid clothing such as neck tie which can get caught in rotating machines or centrifuge.
14. Disconnect power before touching moving parts; use lock out devices.
15. Be careful about the hot plate as one may get injury from it.
16. Wear radiation detection tag while working with radio-isotopes as it will give the information about the level of radiation one has been exposed.
17. Dispose the radio-isotope safely and securely to a place made for it.
18. Do not get exposed to UV light as it may damage the retina and cause skin cancers.
19. Wear gloves while handling the OPD, ODD chromogens as they are potential carcinogens.
20. All discarded plates, tubes, clinical samples and other contaminated materials are to be placed in disposal containers.
21. Special disposal boxes must be used for sharps such as syringes or broken glass disposal.

22. Containers or contaminated materials should be carefully autoclaved before disposal.

23. All areas of the laboratory must be kept clean and tidy and free from dirt, dust etc.

24. Floors should be kept clean and washed with germicidal solution on regular basis and after any spills of infectious material.

25. Refrigerators and freezers should be regularly inspected for the presence of broken vials or tubes containing infectious agents.

26. Refrigerators and freezers should be regularly cleaned and defrosted to prevent contamination and temperature failure.

27. Keep burners away from inflammable materials. Bulk flammable material must be stored in the safety cabinet.

28. Turn off the burners, LPG when not in use.

29. Mouth pipetting should be strictly prohibited in the laboratory. Rubber bulbs should be used.

30. Bench tops should be wiped with a disinfectant (a phenolic disinfectant, 1 % sodium hypochlorite or 70% ethanol), routinely after working with infectious agents or clinical specimens or after spills, splashes or contamination by infectious materials.

31. Gloves should be worn when performing work with potentially hazardous organism or specimens such as HBV, HIV, body fluid, clinical specimens etc.

GLOSSARY

ANEUOPLOID : A cell having a chromosome number different from diploid chromosome number.

Apoptosis : Programmed cell death caused by activation of intracellular degradative enzyme.

Attenuation : It is associated with vaccine. When an organism particularly virus is passaged in the cell/ECE/ animal to make it immunogenic but nonpathogenic, the process is called attenuation.

CELL CULTURE : The growth of cells *in vitro*.

Cell line : Cells that originate by subculture of a primary culture. It may be finite or continuous.

Cell strain : Cells derived from a cell line or primary culture by selection of a specific property or marker which is maintained in subsequent subcultures. The selection of the cell strain is carried out by cloning.

Clone : A population of genetically identical cells derived from an individual cell.

Confluent : The cell sheet cover the entire growth surface in the culture vessel.

Cryopreservation: It is to preserve the cells by freezing at low temperature in the presence of cryoprotective agents *viz.*, glycerol/DMSO.

DIPLOID : When the cell in which chromosomes (except sex chromosome) are paired and structurally identical with those of the species of its origin. Each species has a characteristic number of chromosome (eg human 48).

Diploid cell line : A cell line in which about 75% cells have same karyotype as the normal cells of the species of their origin.

Doubling time: The time taken for a cell population to double in number.

ESTABLISHED CELL LINE : A cell line having the potential to be subcultured indefinitely *in vitro*.

HETEROPLOID CELL LINE : A cell line having < 75% of cells with diploid chromosome numbers.

IMMORTALIZATION : The transformation of a cell with a finite lifespan to a cell with unlimited lifespan.

In vitro : Cell growth outside the body (glass/plastic).

In vivo : Cell growth within the body.

KARYOTYPE: The chromosome complement of an individual cell.

MALIGNANT: A tumour which has the property to invade other tissue (metastatis).

Mitogen : A compound capable of inducing mitosis.

Monoclonal antibody: An antibody which is produced from a hybrid cell synthesizing single antibody type.

Monolayer: The growth of adherent cell on the plastic/glass surface.

Mutant: A variant cell results from an altered gene.

Myeloma: A tumour cell derived from B cells capable of multiplying indefinitely.

NECROSIS: Cell death due to breakdown of plasma membrane leading to cell swelling and rupture.

ONCOGENE : A gene which is implicated in the malignant transformation of cells.

Organ culture : The maintenance or growth of the whole or part of an organ *in vivo*.

PASSAGE NUMBER : The number of subculture performed after the original isolation of the cells from a primary source.

Pluripotent : The capacity of an embryonic cell to differentiate into any cell type.

Polypropylene : It is a type of plastic which can withstand autoclaving.

Polystyrene : It is a type of plastic which can not withstand autoclaving.

Primary culture : A culture of cells taken directly from the tissue of an animal.

SPLIT RATIO : The number of new cultures established from a parent culture during subculture. If cells of one culture flask is used to distribute into 3 new flasks, the split ratio is 1:3.

Substratum : It is the surface upon which cells can attach while growing.

Suspension culture : Cells which can grow while suspended in medium without the need for adherence on the surface.

Syngeneic : Animals which are genetically similar and produced from inbreeding.

TITRE : It is the concentration of infectious agents.

Trypsinization : It is the process in which trypsin is used to remove adherent cells from the surface.

APPENDIX

Coating buffer (Carbonate/bicarbonate buffer, pH 9.6)

Sodium carbonate (Na_2CO_3)	- 1.59 g
Sodium bicarbonate ($NaHCO_3$)	- 2.93 g
Sodium azide	- 0.20g

Made upto 1 litre with distilled water.

PBS-polysorbate 20 (Tween 20), pH 7.4

Sodium chloride	- 8.0 g
Potassium dihydrogen phosphate (KH_2PO_4)	- 0.2 g
Disodium hydrogen orthophosphate (Na_2HPO_4, 2 H_2O)	- 2.9 g
Potassium chloride	- 0.2 g
Sodium azide	- 0.2 g
Polysorbate 20 (Tween 20)	- 0.5 ml
Distilled water up to	- 1 litre

Phosphate citrate buffer (pH 5.0)
Sol. A – 0.1 M Citric acid (1.92 g in 100 ml distilled water).
Sol. B – 0.2 M Na_2HPO_4 (2.84 g in 100 ml distilled water).
24.3 ml of sol. A and 25.7 ml of sol B made upto 100 ml with distilled water.

Substrate solution

Orthophenylene diamine	- 4 mg
Phosphate citrate buffer (pH 5.))	- 10 ml
H_2O_2	- 10 µl

5. Reaction stopping solution

Hank's Medium

Hank's Solution –I

	1(M) H_2SO_4
NaCl	- 80 g
KCl	- 4 g
$MgSO_4, 7H_2O$	- 2 g
$CaCl_2, 2H_2O$	- 1.85 g or $CaCl_2$ - 0.7 g
Distilled water	up to - 1000 ml

Hank's Solution –II

$Na_2HPO_4, 2H_2O$	- 0.75 g
or	
$Na_2HPO_4, 12H_2O$	- 1.5 g
$KH_2PO_4,$	- 0.65 g
Dextrose	- 20 g
Phenol red (0.4%)	- 50 ml
or	
Phenol red (1%)	- 20 ml

Hank's Solution –III

$NaHCO_3$	- 1.4 g
D. Water	- 100 ml

Hanks I and II are autoclaved separately at 15 lbs pressure for 30 min and kept at 4°C till use. Hanks III is to be prepared fresh before adding it to the Hanks medium.

Hank's maintenance medium

Lactalbumin hydrolysate	- 5 g
Yeast extract	- 0.5 g
Hanks I	- 100 ml
Hanks II	- 100 ml
Hanks III	- 25 ml
Penicillin	- 1 lakh units
Streptomycin	- 0.2 g
Distilled water	- upto 1000 ml

The medium is sterilized by passing through EKS filter pads.

Hank's growth medium

Lactalbumin hydrolysate	- 5 g
Yeast extract	- 0.5 g
Hanks I	- 100 ml
Hanks II	- 100 ml
Hanks III	- 25 ml
Serum	- 100 ml
Penicillin	- 1 lakh units
Streptomycin	- 0.2 g
Distilled water	- upto 1000 ml

The medium is sterilized by passing through EKS filter pads.

Phenol red solution (1%)

1 g of phenol red powder dissolved in 100 ml of 0.1 N NaOH. Solution is filtered and stored in the refrigerator.

Glycine buffer, pH 9.5-10.5

Sodium hydroxide	- 5.1 g
Sodium chloride	- 104 g
Glycine	- 156 g
Distilled water	- up to 1000 ml

Alsever's solution

Dextrose	- 29.50 g
Sodium chloride	- 4.20 g

Sodium citrate - 8.00 g
Citric acid - 0.55 g
Distilled water - up to 1000 ml

Phosphate buffer saline (PBS)
PBS (Solution A)
NaCl - 64 g
KCl - 1.6 g
$Na_2HPO_4\ 7H_2O$ - 17.36 g
KH_2PO_4 - 1.6 g
Distilled water - 6400 ml
PBS (Solution B)
Calcium chloride - 0.8 g
Distilled water - 800 ml
PBS (Solution C)
Magnesium chloride ($MgCl_2$, $6\ H_2O$ - 1.7 g)
Distilled water - 800 ml

All the solutions (A, B and C) are sterilized separately at 20 lbs at 115°C for 30 minutes in an autoclave.

Final preparation of Phosphate Buffer Solution (PBS)

After cooling, add solution B to A and then C to BA. 1 lakh I.U. of crystalline penicillin and 100 mg of dihydro-streptomycin per litre of PBS are added with sterile precautions. This is finally distributed in smaller flasks of 100, 250 and 500 ml capacities and stored in refrigerator.

Trypsin-versene solution

NaCl = 10 g
KCl = 0.250 g
Na_2HPO4 = 1.90 g
KH_2PO4 = 0.25 g
Trypsin = 1.7 g
EDTA = 1.4 g
0.4% Phenol red = 1 ml
Distilled water = 1000 ml

Mix the contents by putting it on a magnetic stirrer and filter it through disposable filter (0.22 µM) and dispensed in sterilized container and keep at -20°C until further use.

Trypan blue (0.4%) for viable counts

Trypan blue stain	= 0.2 g
Normal saline solution	= 50 ml

Dissolve the trypan blue in 50 ml of NSS. Filter and aliquot in 5-10 ml quantity. Store at room temperature. It is also available commercially as 0.4% trypan blue solution.

50% buffered glycerine saline

PBS (pH 7.4)	= 50 parts
Glycerine	= 50 parts

10% Formol saline

Formaldehyde (40%)	= 100 ml
Sodium chloride	= 8.9 g
Distilled water	= 900 ml

Virus transport medium

Two common virus transport media used are (1) 50% glycerine phosphate buffer saline and (2) Hank's balanced salt solution (HBSS) containing 0.5% bovine serum albumin or 1% FCS and antibiotics.

BSA fraction V	= 5 g
or FCS	= 10 ml
Penicillin	= 10 lakhs units
Streptomycin	= 1 g
Mycostatin	= 25,000 units
HBSS	= up to 1000 ml

Adjust the pH 7.0 and sterilize by membrane filtration and aliquot into screw capped glass or plastic vials. As soon as the specimen is collected, it should be maintained at an optimum pH (7.2 & 7.4) and should be transported on ice to the laboratory.

REFERENCES

Barnes, D., Sirbsku, D. and Sato, G. (1994). Cell Culture : Methods for Molecular and Cell Biology, 4 volumes, Wiley-Liss, New York.

Burleson, F.G., Chambers, T.M. and Wiedbrauk, D.L. (1992). Virology a Laboratory Manual. Academic Press.

Butler, M. (1987). Animal Cell Technology : Principles and Products, Open University Press, New York.

Butler, M.J. (1997). Animal Cell Culture and Technology, IRL Press, Oxford, U.K.

Darling, D.C. and Morgan, S.J. (1994). Animal Cell Culture. Bios Scientific Publishing Ltd. Wiley, New York.

Darling, D.C. and Morgan, S.J. (1994). Animal Cells : Culture and Media. Wiley, New York.

Davis, J.M. (1994). Basic Cell Culture : A Practical Approach. IRL Press, Oxford, U.K.

Doyle, D., Hay, R. and Kirsop, B.E. (1991). Animal Cells : Living resources for Biotechnology. Cambridge University Press, Cambridge, U.K.

Freshney, R.I. (1992). Animal Cell Culture : A Practical Approach, IRL Press, Oxford.

Harrison, M.A. and Rac, I.F. (1997). General Techniques of Cell Culture, Cambridge University Press, Cambridge, U.K.

Masters, J.R.W. (2000). Animal Cell Culture. Oxford University Press, Oxford, U.K.

Mather, J.P. and Barnes, D. (2006). Animal Cell Culture Methods, Academic Press.

Paul, J. (1973). Cell and Tissue Culture. Churchill Livingtone, Edinburgh and London.

Pollard, J.W. and Walker, J.M. (1990). Animal Cell Culture : Methods in Molecular Biology, Vol 5, Humana Press.

Sasakim R. and Ikura, K. (1991). Animal Cell Culture and Production of Biologicals. Kluwer, Boston, MA.

Wasley, J.D. and May, J.W. (1971). Animal Cell Culture Methods, Lippincott-Raven Publishers.

CPSIA information can be obtained
at www.ICGtesting.com
Printed in the USA
LVHW061523250619
622311LV00003B/18/P